A TECHNICAL ECONOMIC MODEL FOR INTEGRATED WATER RESOURCES MANAGEMENT IN TOURISM DEPENDENT ARID COASTAL REGIONS; THE CASE OF SHARM EL SHEIKH, EGYPT

A Technical Economic Model for Integrated Water Resources Management in Tourism Dependent Arid Coastal Regions; the Case of Sharm El Sheikh, Egypt

DISSERTATION
Submitted in fulfilment of the requirements of
the Board for Doctorates of Delft University of Technology
and of the Academic Board of the UNESCO-IHE Institute for Water
Education for the Degree of DOCTOR
to be defended in public
on Tuesday, October 13, 2009 at 10:00 hours
in Delft, the Netherlands

by

Aya Lamei
born in Cairo, Egypt

Master of Science in Environmental Engineering
The American University in Cairo, Egypt

CRC Press is an imprint of the
Taylor & Francis Group, an **informa** business
A BALKEMA BOOK

This dissertation has been approved by the supervisors:

Prof. dr. ir. P. van der Zaag	UNESCO-IHE/TU Delft, The Netherlands
Prof. dr. E. Hamdy Imam	The American University in Cairo, Egypt

Members of the Awarding Committee:

Chairman:	Rector Magnificus, TU Delft, The Netherlands
Vice-chairman:	Rector, UNESCO-IHE, The Netherlands
Prof. dr. ir. P. van der Zaag	UNESCO-IHE/TU Delft, The Netherlands
Prof. dr. E. Hamdy Imam	The American University in Cairo, Egypt
Prof. dr. M. Balaban	MIT, USA
Prof. dr. M.P. van Dijk	UNESCO-IHE and ISS, The Netherlands
Prof. ir. J.C. van Dijk	TU Delft, Netherlands
Prof. dr. ir. H.H.G. Savenije	TU Delft, Netherlands
Prof. dr. ir. D. Brdjanovic (reserve)	UNESCO-IHE/TU Delft, The Netherlands

CRC Press
Taylor & Francis Group
6000 Broken Sound Parkway NW, Suite 300
Boca Raton, FL 33487-2742

First issued in hardback 2017

Published by:
CRC Press/Balkema
PO Box 447, 2300 AK Leiden, The Netherlands
e-mail: Pub.NL@taylorandfrancis.com
www.crcpress.com – www.taylorandfrancis.co.uk – www.balkema.nl

ISBN-13: 978-0-415-55898-3 (pbk)
ISBN-13: 978-1-138-43399-1 (hbk)

Visit the Taylor & Francis Web site at http://www.taylorandfrancis.com

and the CRC Press Web site at http://www.crcpress.com

Table of Contents

List of Symbols

a	annuity factor used to calculate amortized costs
A	open surface area of an evaporation pond (m^2)
a_b	beach access lost due to water quality degradation (km)
A_{gr}	area of irrigated land (m^2)
A_p	area of swimming pool (m^2)
b	constant
B	present value of the net benefit over the planning horizon of the project (US$)
c	unit cost of different O&M components for a reverse osmosis desalination plant (US$/$m^3$)
C	net present value of user or external cost (US$/$m^3$)
C_c	unit capital cost of water supply (US$/$m^3$/d)
c_{ch}	unit cost of chemicals for a reverse osmosis desalination plant (US$/$m^3$)
$C_{D,LR,R}$	cost of damage or lost revenue or cost of replacement technology (US$/$m^3$)
C_E	direct cost of existing technology (US$/$m^3$)
c_e	unit cost of energy for a reverse osmosis desalination plant (US$/kWh)
C_{eff}	water demand conservation effectiveness expressed as a percentage
c_l	unit cost of labour and maintenance for a reverse osmosis desalination plant (US$/$m^3$)
C_{LR}	lost revenue per cubic meter of desalinated water for loss of beach access (US$/$m^3$)
c_m	unit cost of maintenance for a reverse osmosis desalination plant (US$/$m^3$)
C_p	unit production cost of water supply (US$/$m^3$)
$C_{T,y}$	Minimum total cost of domestic and irrigation water in year y (US$/yr)
C_y	present value of the cost of water production in year y (US$/yr)
d	constant
DC	direct cost of a project (US$/$m^3$)
D_y	domestic demand at year y (equivalent to contracted-for water supply or basic demand) (m^3/yr)
D_Y	maximum demand reached at the end of the planning horizon of the project summing up both contracted-for supply and extra potable water demand (m^3/yr)
e	constant
e_f	fixed water consumption (m^3/d)
EC	external cost of a project (US$/$m^3$)
e_v	evaporation rate (m/d)
f	safety factor
f_{dw}	factor indicating the percentage of domestic water demand ending up as sewage (%)
f_e	factor for specific energy consumption (kWh/m^3)
f_g	fraction for number of guests per room
f_o	fraction of the peak demand
f_s	fraction for number of staff per room
f_{un}	factor for unaccounted-for water (%)
f_{wdm}	factor for water demand management level (%)

F_Y	maximum present value of total net benefit over the planning horizon of the project (US$)
i	discount rate (%)
L	length of water supply pipeline (km)
l	length of shoreline (km)
n	economic plant life (yr)
N_g	number of guests
N_r	number of rooms
N_s	number of staff
O_j	average monthly occupancy rate in month j (%)
$O_{ac,j}$	actual occupancy rate in month j (%)
p	selling price of energy (US$/kWh)
P_{EP}	selling price of potable water in excess of contracted-for water supply (US$/m^3)
P_{EWW}	selling price of treated wastewater in excess of contracted-for waste water supply (US$/m^3)
P_p	selling price of contracted-for potable water (US$/m^3)
P_R	price of replacement technology (US$/m^3)
P_{TWW}	unit cost of treating wastewater within the premises of the hotel (US$/m^3)
P_{WW}	selling price of treated wastewater bought from an external source (US$/m^3)
Q	total water consumption of the entire beach area (m^3/yr)
$Q_{ac,j}$	actual water demand in month j (m^3/d)
$Q_{c,j}$	daily contracted-for water supply in month j (m^3/d)
$Q_{dom,j}$	daily domestic water demand of a hotel in month j (m^3/d)
$Q_{ec,j}$	required potable water supply in excess of contracted-for water supply in month j (m^3/d)
Q_{ed}	excess demand due to difference between expected actual and average occupancy rate in month j (m^3/d)
$Q_{I,j}$	daily water demand of irrigation in month j (m^3/d)
$Q_{IC,j}$	contracted-for irrigation water supply in month j (m^3/d)
$Q_{IEC,j}$	irrigation water demand in excess of contracted-for irrigation water supply in month j (m^3/d)
$Q_{Ipeak,j}$	peak irrigation water demand of a hotel in month j (m^3/d)
$Q_{peak,j}$	peak domestic water demand of a hotel in month j (m^3/d)
Q_r	volume of reject brine from a reverse osmosis desalination plant (m^3/d)
$Q_{s,j}$	water supply from a reverse osmosis desalination plant in month j (m^3/d)
Q_w	capacity of water supply (long-distance piping or desalination) expressed as flow rate of product water (m^3/d)
Q_{ww}	daily wastewater flow rate (m^3/d)
r	monthly growth rate in water consumption (month^{-1})
R_y	present value of expected revenue from water sales in year y (US$/yr)
SC_g	specific water consumption of guests (m^3/cap/d)
SC_{grj}	specific water consumption of irrigated land in month j (m^3/m^2/d)
$SC_{p,j}$	specific water consumption of swimming pool in month j (m^3/m^2/d)
SC_r	specific water consumption of hotel room including all side activities (m^3/room/d)
SC_s	specific water consumption of staff (m^3/cap/d)
SC_{sh}	specific water consumption of staff housing (m^3/cap/d)
S_y	possible states or reverse osmosis plant capacity of the system at different stages of the project (m^3/d)
T	average yearly temperature (°C)
TC	opportunity cost measured in economic terms (US$/m^3)

TR	total expected yearly revenue from an activity (US$/yr)
U	utilization factor (%)
UC	user cost of a project (US$/m^3)
v	constant
V_y	cost or revenue in year y (US$/yr)
V_o	present worth of cost or revenue (US$/yr)
w	constant
y	number of years
Y	planning horizon of the project (yr)
x_y	reverse osmosis plant capacity expansion at different stages of the project (m^3/d)
z	constant

1 General Introduction

Introduction

Water scarcity aggravates in coastal zones which are characterized by high population density, intense economic activity and tourism, and consequently heavy seasonal water demand. The usual way to tackle water demand is through conventional surface and ground water abstraction. However, due to increasing limitation in water resources, a shift is taking place towards integrated water resources management (IWRM).

IWRM adopts a holistic approach to optimize water usage. IWRM has to take into account the following four dimensions (Savenije and van der Zaag, 2008):

- Water resources including stocks and flows as well as water quantity and quality. It critically assesses supply options, including developing alternative water resources, e.g. desalination (removal of dissolved minerals including salts) and reclaimed wastewater. It also seeks to increase the management efficiency of conventional resources and schemes and will also consider demand management options.
- Water users, considering all sectoral interests and stakeholders, including the environment and future generations.
- Spatial dimension, including the spatial distribution of water resources and water demands, and the various spatial scales at which water is being managed.
- Temporal dimension, considering the temporal variability in availability of and demand for water resources and the physical structures that have been built to even out fluctuations and to better match supply with demand.

IWRM projects should be sustainable and fulfil the public interest. Sustainability should be considered from the following perspectives:

- maintenance of environmental quality (including water quality)
- financial sustainability (cost recovery)
- good governance (effective management mechanism)
- institutional capacities (capacity building, human resources, and appropriate policy and legal framework)
- social equity (equal right of people to water resources)

However, the development, decision making process and implementation of projects conforming to the IWRM approach are complex due to the different sectors involved, typically water, environment, economy, energy and agriculture (see Figure 1.1) (Thomas and Durham, 2003).

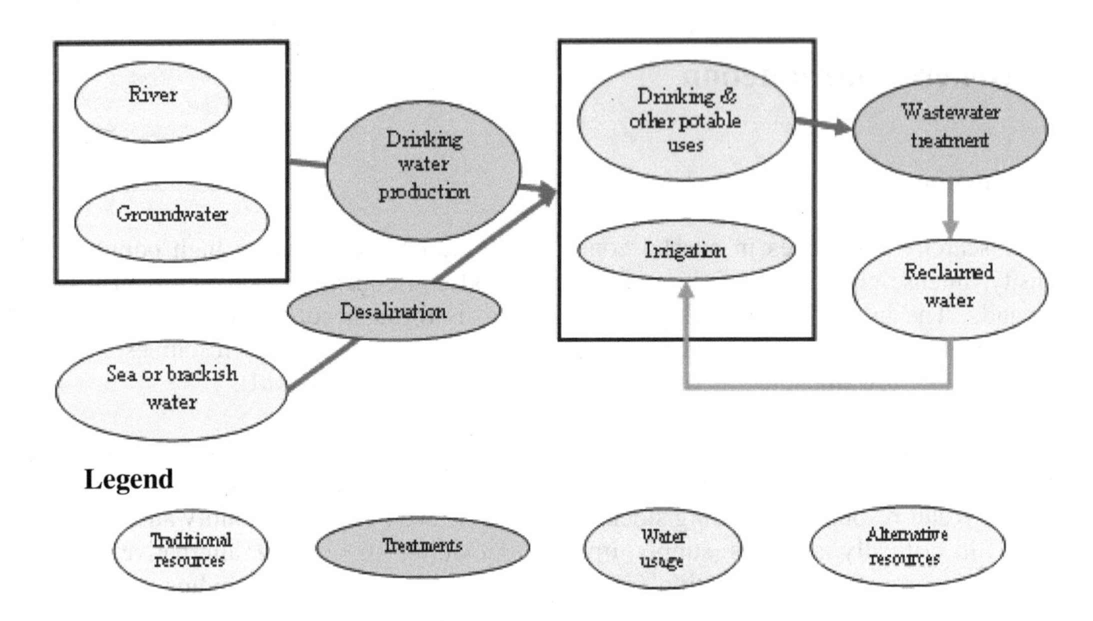

Figure 1.1 Integrated water resources management approach (adapted from (Thomas and Durham, 2003))

A technical-economic modeling tool is developed for public and private decision makers. The decision maker using the modeling tool can be a hotel, a water company, or a water management utility (i.e. municipality, city, town, and governorate). The water management utility is assumed to perform integrated water management including: supply of potable water, collection and treatment of wastewater, and reuse of treated wastewater for irrigation water supply.

Egypt is among those countries which are very vulnerable with regard to water resources. It is located in a belt of extreme aridity being the country furthest downstream in the Nile basin. The pressure of water scarcity, for regions located along the length of the Nile, is already mounting. A more disadvantaged region is the Red Sea coastal area, where fresh water is not always available.

In South Sinai, along the Red Sea, tourism is the dominating industry. Water is sourced by either desalination of sea or brackish water, or by piped or trucked water from the Nile (Abou Rayan *et al.*, 2001). Despite the water scarcity, reuse of treated wastewater is not widely applied. Treated wastewater could be used for irrigation and non-potable domestic use, thereby reducing potable water demand. Due to costs of desalination and water transportation, the price of water in this tourist region is about ten times higher than in Cairo. What further contributes to price escalation is the dependence on privately-owned small size desalination plants for water supply, i.e. no economies of scale.

This region is facing continuous economic growth. The problem is that economic development is not matched by adequate water resources. The gap between water demand and water availability is expected to reach 1 Mm^3/d (million cubic meter per day) in South Sinai by the year 2020 (Hafez and El Manharawy, 2002). Due to limited Nile resources and excessive transportation costs, the scenario for development of the region has to be based on seawater desalination and reuse of treated wastewater to satisfy all demands, as well as water demand management (decrease water demand and use).

The private sector must be encouraged to invest in water projects to relieve increasing financial pressures on the government, and to satisfy increasing water demand. A detailed study of the situation with modeling of possible management scenarios for projects will aid investors (both public and private) in the decision making process (Khalil, 2004).

The research problem

Tourism-dominated arid coastal regions have special characteristics which have to be considered when planning for integrated water resources management. They have limited water resources, while having to deal with a highly fluctuating water demand and consequent regular water shortage. Increased water demand leads to increased water prices and limited planning time which can eventually degrade the sensitive environment impacting the tourism industry. IWRM projects have to consider both financial and environmental sustainability ensuring cost recovery from projects (both public and private) and minimizing the impact on the environment.

Since conventional surface and groundwater resources are limited, water can be sourced by either reverse osmosis (RO) desalination of sea or brackish water, or by long-distance tanker trucks or a pipeline (long-distance defined here as equal to or further than 30 km). Though RO desalination can be comparable and even cheaper to long-distance piping, RO desalination can have some negative environmental impact due to high energy consumption per m^3 of fresh water produced and brine containing high salt concentrations and chemicals. A small change in energy prices (measured in US$/kWh) directly affects the unit production costs of RO plants.

Hotels in arid coastal areas use mainly desalinated water for their domestic water demands, and treated wastewater for irrigating green areas (excess irrigation demand is satisfied by supplying desalinated water at a much higher cost). Private water companies supply these hotels with their domestic and irrigation water needs. There is normally a contractual agreement stating a minimum requirement that has to be supplied by the water company and that the hotel has to pay for regardless of its actual consumption ("contracted-for water supply").

Figure 1.2 illustrates the setup investigated in this research. A hotel should contract wisely in order to minimize its total annual water costs and ensure adequate water supply to the facility for both domestic and irrigation water demand.

From the supply side, a water company establishing a RO desalination plant designed to satisfy only the contracted-for water demand would be missing out on potential benefits that could have been obtained selling water in periods of high demand. On the other hand, sizing the RO desalination plant to produce water to satisfy the peak demand means incurring additional costs as well as having the plant partially idle during periods of average or low demand. Moreover, future change in demand has to be taken into consideration for optimum capacity expansion installation.

Having an on-site wastewater treatment plant presents savings to the hotel. A conscious strategy for water management should rely solely on treated wastewater on-site for non-potable uses, requiring careful planning of the type of plantation and the size of green area.

The principal research questions of this PhD research are:

- What is the current setup for water resources management projects in tourism-dominated arid coastal regions, with specific reference to Sharm El Sheikh? What are the shortfalls and how can these be addressed in order to achieve financial and environmental sustainability?
- Which potable and non-potable water supply options are relevant to tourism dominated arid coastal regions?
- What are the costs associated with the selected water supply options?
- How does the cost of energy influence the cost of RO desalinated water? Can renewable sources of energy be an option?
- What are the factors influencing water demand in a hotel?
- What is the optimum capacity of an RO desalination plant that ensures continuous supply and minimizes water production costs?
- Given the variability in water demand by a hotel, what would be the optimal contractual agreement for both potable and non-potable water supply?
- What is the size of a green area of a hotel that can minimize dependence on external sources of water supply and be sustained by available grey water from the hotel? How can irrigation water demand be minimized for a hotel?
- What are the selection criteria for the brine disposal method from an RO plant taking into consideration cost and environmental impact?

Scope, objectives and relevance

The study's scope is to develop a technical-economic modeling tool to aid decision makers (both public and private sector investors) in the design and assessment of integrated water resources management projects for arid coastal regions. Designed projects are to satisfy varying temporal and spatial demand and to find sustainable solutions (which may include a higher level of wastewater reuse). The evaluation of the IWRM projects will be based on a cost benefit analysis and environmental impact.

Objective 1: Perform a literature review for practiced IWRM and available models
Survey available models targeting IWRM projects, as well as practiced options for achieving IWRM in different arid coastal regions worldwide, in order to match water availability to water consumption now and in the future in a sustainable manner.

Objective 2: Develop a technical-economic modeling tool to invest in sustainable IWRM projects in tourism-dominated arid coastal regions composed of the following models: water demand, water supply, reclaimed water, by-products and an economic model
Develop a technical-economic modeling tool specially designed for sustainable IWRM projects in tourism-dominated arid coastal regions. The modeling tool presented in this research is composed of several models: time-variant water demand, water supply, wastewater reuse, environmental (by-products) and economic models. The modeling tool is developed using Excel worksheets.

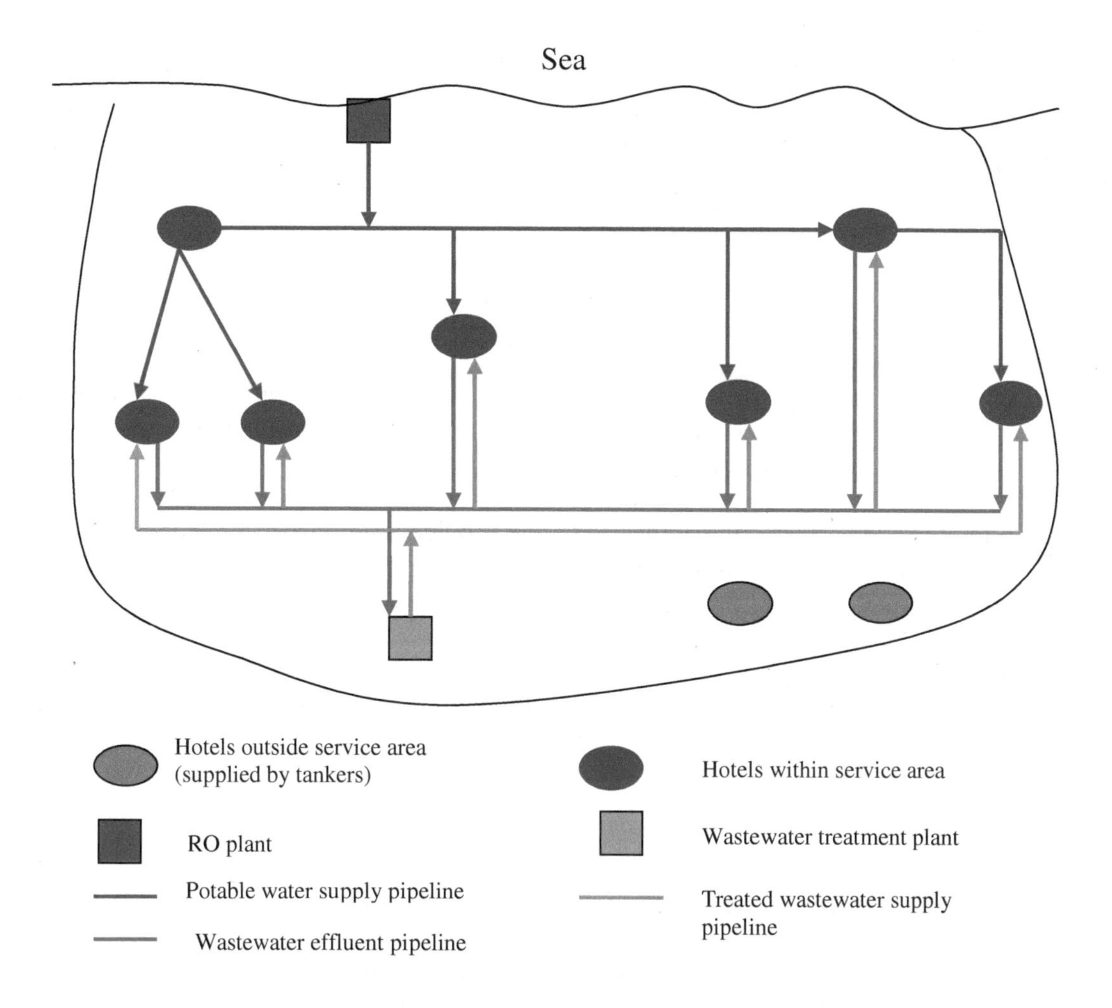

Figure 1.2 The physical setup of IWRM projects in the case study area including water supply and wastewater treatment plants and users

Objective 3: Application and testing of the technical-economic modeling tool on a case study

Demonstrate the feasibility of the developed technical-economic modeling tool for sustainable IWRM projects in the case study region (Sharm El Sheikh, South-Sinai, Egypt). Several design and operation scenarios are compared based on a cost-benefit analysis and environmental impact (where applicable). The scenarios address the following issues:

1. Capacity of desalination plants/ wastewater treatment plants serving a combination of demand nodes
2. Desalination and wastewater treatment technologies
3. Demand management options
4. Alternatives to desalination, e.g. Nile water
5. Energy sources including renewable energy
6. Alternative disposal methods of brine from desalination and their environmental impact

In order to test the developed technical-economic modeling tool, a set of data has been collected to be used as input variables. The data include:

- a local water balance

- specific water and energy consumption rates
- costs for water supply, wastewater treatment and reuse, brine disposal and energy

Relevance

The relevance of the research is as follows:

1. The study will generate information that may guide investments in sustainable water projects;

2. The study diverts from the conventional notion of dealing with wastewater as a non-valuable resource. This will ameliorate the environmental conditions in sensitive arid coastal zones;

3. Implementing the results of this study will provide a solid basis for economic and sustainable development in this area ensuring adequate water resources;

4. Better and more efficient management of water resources in the area will lead to better service, lower gross water use levels and lower prices, positively impacting the tourism sector, as well as the communities that directly depend on this industry.

Literature review

Water resources in arid coastal regions

Water resources in arid coastal regions are characterized by special features attributed to geographic location and a sensitive environment. In addition, in many of these regions, tourism is the dominating economic activity. These special features have a unique effect on water demand, available water supplies, wastewater generation, choice of treatment technology, and disposal of by-products.

Water demand
Water demand is periodic in these regions depending on tourists' occupancy rates which vary according to time of the year and other economical/ marketing variations. Also, the tourism industry requires huge amounts of water (between 300 and 850 l/cap/d, much more than a local inhabitant requires (Chartzoulakis *et al.*, 2001; Gonzalez *et al.*, 2005), and tourists are normally not prepared to encounter water scarcity.

Froukh (2001) developed a decision support system for forecasting the average household consumption based on water consuming devices used, socio-economic and demographic factors. Another model was developed mainly using past trends to forecast present domestic water demands (Ramjeawon, 1994).

In this PhD research, past trends in occupancy rates and specific water consumption rates are used to forecast future demands; different scenarios are tested in order to take into consideration possible variations.

Water supply
Conventional water supplies are surface and ground water. However, in many cases, conventional sources are either limited or not evenly distributed where it is needed. Non-conventional sources include seawater and brackish water desalination and reclaimed wastewater. Other possibilities could include transportation of water by pipes, trucks, railways or ships (Salgot and Tapias, 2004). Many countries are considering water supply augmentation through non-conventional resources in addition to policy remedies encouraging domestic and irrigation water demand management (Al Zubari, 2003).

Water supply options for the case study area are investigated, in this research, in terms of availability, energy demand, cost and impact on the environment. Environmental limitation associated with desalinated water is the production of brine. A usual disposal medium is the sea. However, there is a growing concern about the effect of brine on the marine ecosystem (Salgot and Tapias, 2004). Chemical composition of brine, temperature, among other factors can cause damage to marine flora and fauna in the vicinity of the brine outlet pipe. Applicable water demand management options are discussed in the research.

Wastewater reuse
In coastal regions, sea disposal of wastewater should not be assumed to be the only option. The damage to sensitive marine ecosystems can be significant. Therefore, reuse of wastewater is a complementary solution and not just an alternative (Paraskevas *et al.*, 2004).

In the process of implementing wastewater reuse projects, decision factors such as wastewater treatment technology, type of reuse, and capacity of plant have to be determined. Conventional wastewater treatment technology includes primary treatment (grit removal, screening, and sedimentation); and secondary treatment which enables removal of biodegradable organic matter and suspended solids. For wastewater reuse, tertiary treatment has to be employed.

Tertiary treatment may be defined as any treatment process in which unit operations are added to the flow scheme following conventional secondary treatment. (Sonune and Ghate, 2004). The use of different technologies has to be evaluated according to the type of treatment needed, costs associated, availability, and the added value produced with this reclaimed water.

Flow of wastewater is directly related to the season and tourists' occupancy rate, i.e. high production rate of wastewater is expected in summer and during periods of high tourists' occupation rate, and vice versa. This temporal variation has to be considered in determining the plant capacity.

Two major types of wastewater reuse are being practiced around the world: 1) potable uses, which can be direct, after high levels of treatment, or indirect, after passing through the natural environment and 2) direct or indirect non-potable uses in agriculture, industry and urban settlements (Lazarova *et al.*, 2001). The focus of this research is on wastewater reuse for irrigation which is the most applicable reuse option for the case study region in order to provide a supplementary source of water instead of using expensive desalinated water for irrigation. Shortage of water resources and increased water demand from local population and tourists especially in summer months makes reuse of wastewater the only option for landscape irrigation in coastal tourist regions (Barbagallo *et al.*, 2003; Shelef and Azov, 1996; Tanik *et al.*, 2005).

Golf course irrigation is the fastest-growing reuse application in arid and semi-arid regions (e.g. Egypt, Tunisia, Morocco) because of its high rate of water consumption (Bahri *et al.*, 2001; Lazarova *et al.*, 2001). The feasibility of golf courses in terms of water availability and associated costs is analyzed in this research.

Practiced water resources management

Water resources management in arid coastal regions ranges from water demand management, to the development of alternative water resources and their integration into the water resources system.

Water demand management
An example of a water demand management (WDM) project addressing the tourism industry was implemented in several resorts in arid Namibia. WDM is here defined as a management approach that aims to "decrease water demand by promoting efficient water use through economic, educational and technological means" (Van der Merw, 1999). The WDM project included awareness programs for visitors and staff, tight maintenance schedules, water pricing for visitors and staff members, landscaping of gardens and retrofitting water outlets with water efficient devices (Schachtschneider, 2000).

Integrated water resources management
Some examples are presented in this section for IWRM projects. In the Canary Islands, wastewater reuse has increased significantly. The treated wastewater is used for agricultural irrigation reducing demand on potable water.

In Hawaii, another IWRM project was implemented. The project included a tertiary treatment system for the production of unrestricted reuse water for irrigation, and a distribution network (Durham *et al.*, 2003). Integrated water management including water supply, sewerage and drainage was introduced as part of the water reform initiative in New South Wales instead of managing the different systems as separate entities (Anderson and Iyaduri, 2003).

The present PhD research discusses the operation of desalination (including disposal) and wastewater plants as part of an integrated water resources management approach.

Modeling of integrated water resources management

Few mathematical models have been developed to address integrated water resources management in arid regions. Other models have only tackled water supplies.

IWRM modeling
A number of models are found in the literature which are presented in this section. Gonzalez *et al.* (2005) developed a decision aid tool (DAT) to assess utilization of groundwater and wastewater reuse (for aquifer recharge) in the water supply of seasonally-stressed regions. The DAT mainly looked at alternative recharge/extraction scenarios based on economic and environmental criteria.

Xu *et al.* (2003b) developed an integrated technical-economic modeling framework to help planning and managing of water resources in a Mediterranean tourist area, in France and Spain. Four types of models were established and coupled: hydrological model, water demand and/or need model; reclaimed water storage model; and a technical-economic model. In addition, a multi-criteria analysis was utilized for the evaluation of scenarios. The target of their model was to integrate the reuse of treated wastewater for irrigation reducing demand on potable water. The authors focused on developing a modeling tool to facilitate the implementation of water reuse in water resources management. Beside the integrated model developed by Xu *et al.* (2003a), other models were designed for integrating wastewater reuse.

Oron (1996) developed a management model for optimal wastewater treatment, disposal and reuse. The model took into account the choice of treatment technology, reclaimed water supply and demand, transportation and storage requirements (for quantity), expected costs from environmental control, and return from selling by-products to treatment process.

Another model developed by Brimberg *et al.* (1993) aimed at developing marginal water resources for arid regions. The marginal sources included saline groundwater, reclaimed water and runoff water. The authors recommended wide application of reclaimed water owing to its economics.

A decision support system (DSS) was developed by Ahmed *et al.* (2002) for the selection of an optimum water reuse scheme of agricultural drainage water. The DSS compares options for treatment processes including desalination, for different use applications. The evaluation was based on technical and economic considerations. The output is the selected treatment process combination, reclaimed water characteristics, and associated costs. The main components of the DSS were: database, user interface, and an expert system incorporating a mathematical model.

Another DSS (WaterGuide) was developed by Loetscher (2000) to match available treated wastewater to demand for non-potable water. The purpose of WaterGuide was to support planners and other stakeholder groups during the early stages of a water recycling project by helping them identify suitable alternatives examining costs, treatment, risks and merits, and by facilitating the assessment of these alternatives with regards to stakeholder preferences.

The technical-economic modeling tool developed in the present research will follow an integrated approach including not only reuse of wastewater but also sources of potable water supply, optimization of contracted-for water supply, required plant capacities expansion, and brine disposal options.

Comparison between previous work and the present research

This section presents a comparison between available models on integrated water resources management and the present research. The most comprehensive published research, that was found, is the technical economic model by Xu *et al.* (2003b). Yet, the authors did not address in detail the variation in potable water supplies, e.g. increasing the capacity of desalination plants, desalination technology and brine production and disposal. The aim of their study was to reduce dependence on desalination water. However, for many arid coastal regions, desalination is the main source of water while other sources are either not available or even more expensive. Therefore, in addition to addressing the implementation of wastewater reuse, economic and environmental consequences of desalination expansion have to be considered including optimum capacity, expansion schedule, brine disposal, and associated costs.

Also, in the study performed by Xu *et al.* (2003a), tourists constituted only 2% of the total water demand. In the present research, the focus is on regions dependent on tourism (more than 50% of the total water demand), which means that monthly and daily fluctuations in water demand are very high. Consequently, generation of wastewater effluent is highly time-variant. Modeling is needed to simulate water demand and wastewater effluent in order to determine optimum capacity of water supply and utilization of reclaimed water.

Several studies have discussed the reuse of wastewater and its integration into the water resources system (Xu, 2002). However, the overall integrated approach of using treated wastewater along with desalination has not been fully investigated. For regions which highly depend on desalination, this integration by use of a mathematical model is inevitable (Khalil, 2004).

A technical-economic modeling tool enables the user to consider many factors at the same time, making it possible to arrive at an economic and environmentally sensible decision. Different IWRM models have been developed. However, arid coastal regions have specific characteristics requiring adapted models to adequately address the challenges associated with IWRM in these regions (Voivontas *et al.*, 2003).

Description of the technical-economic modeling tool

Time-variant water demand model
A time-variant water demand model for domestic population, hotels, and agriculture (landscape) estimates present and future water consumption. The model calculates average and peak consumption. Peak consumption accounts for seasonal and daily variations. Average and peak consumption is used to determine average and peak water demand (taking into consideration losses in water distribution and treatment). The water demand model considers the water demand of different users: guests, pool, staff, and irrigated green area (Chapters 5 and 6).

It should be pointed out that all water uses, except irrigation, requires water of potable quality. Irrigation water can be treated wastewater. Using CROPWAT4, irrigation water demand is estimated on a daily basis taking into consideration irrigation efficiency, type of cultivation, and local weather conditions.

Water supply model
The water supply model has input variables including alternative water sources, required water demand, quality, and sources of energy. The water demand model feeds into the water supply model to determine required capacity of the desalination plant, taking into consideration availability of alternative water sources.

The water supply model calculates the quantity of contracted-for water which is needed from the water company and the amount of extra water needed (in excess of contracted-for water supply). Excess water can be supplied by the water company itself (depending on availability) and/or from private individual trucks (at higher water prices than from the water company) (Chapter 6). The water supply model compares two options of water supply: desalination and long-distance piping based on a cost benefit analysis (Chapter 3). Since, desalination requires substantial amounts of energy; a research was done to investigate the possibility of using solar energy instead of conventional fossil fuel (Chapter 4). Dynamic programming is used in the water supply model to optimize the capacity expansion schedule of RO desalination plants (Chapter 7).

Reclaimed water model
The reclaimed water model uses the potable water flux from the water supply model as an input to calculate wastewater flows. The flow of wastewater is used to calculate required capacity of plant and optimize contracted-for irrigation water supply with the objective function to minimize the overall cost of domestic and irrigation water supply to hotels (Chapter 8). The model calculates the size of the irrigated area which can be sustained by the available grey water treated on-site.

By-products disposal and treatment model
The by-products disposal and treatment model calculates the amount of brine produced from desalination plants, estimates the impact on marine system in case of sea disposal, and compares different disposal methods with respect to their total cost (i.e. direct and indirect costs) (Chapter 9).

Economic model
The economic model calculates costs and benefits associated with the overall water resources project including water supply, reclaimed water, energy, transportation, and brine disposal. Alternatives are analyzed and evaluated. The cost model calculates the cost of water to hotels in periods of high and low demand: cost of contracted-for water supply, cost of excess water if needed, cost of treating wastewater on-site, and cost of buying treated wastewater from an external source. The economic model is interlinked with each of the other models.

Table 1.1 illustrates the different models with their input and output. Figure 1.3 illustrates the main components of the developed technical-economic modeling tool and the coupling of the mathematical models.

Table 1.1 The different models with their input and output

Model	Input	Output
Water demand	Population, hotel occupancy rates, number of tourists, specific consumption rates, loss factor, agricultural consumption, etc.	Current and future domestic, tourist and agricultural average and peak water demands
Water supply	Alternative water sources, transportation costs, water demand, energy sources, prices	Preferred water supply source, Potable water flux, optimum contracted-for potable water supply, optimized capacity, expansion strategy
Reclaimed water	Potable water flux, wastewater treatment technologies, prices	Reclaimed water flux, capacity, optimum contracted-for irrigation water supply
By-products disposal and treatment	Flux of brine produced, possible disposal methods, prices	Selected disposal/treatment method for brine, direct and indirect costs
Economic	Prices, Capital costs, operation and maintenance costs, energy costs	Cost/ benefit analysis, evaluation of alternatives

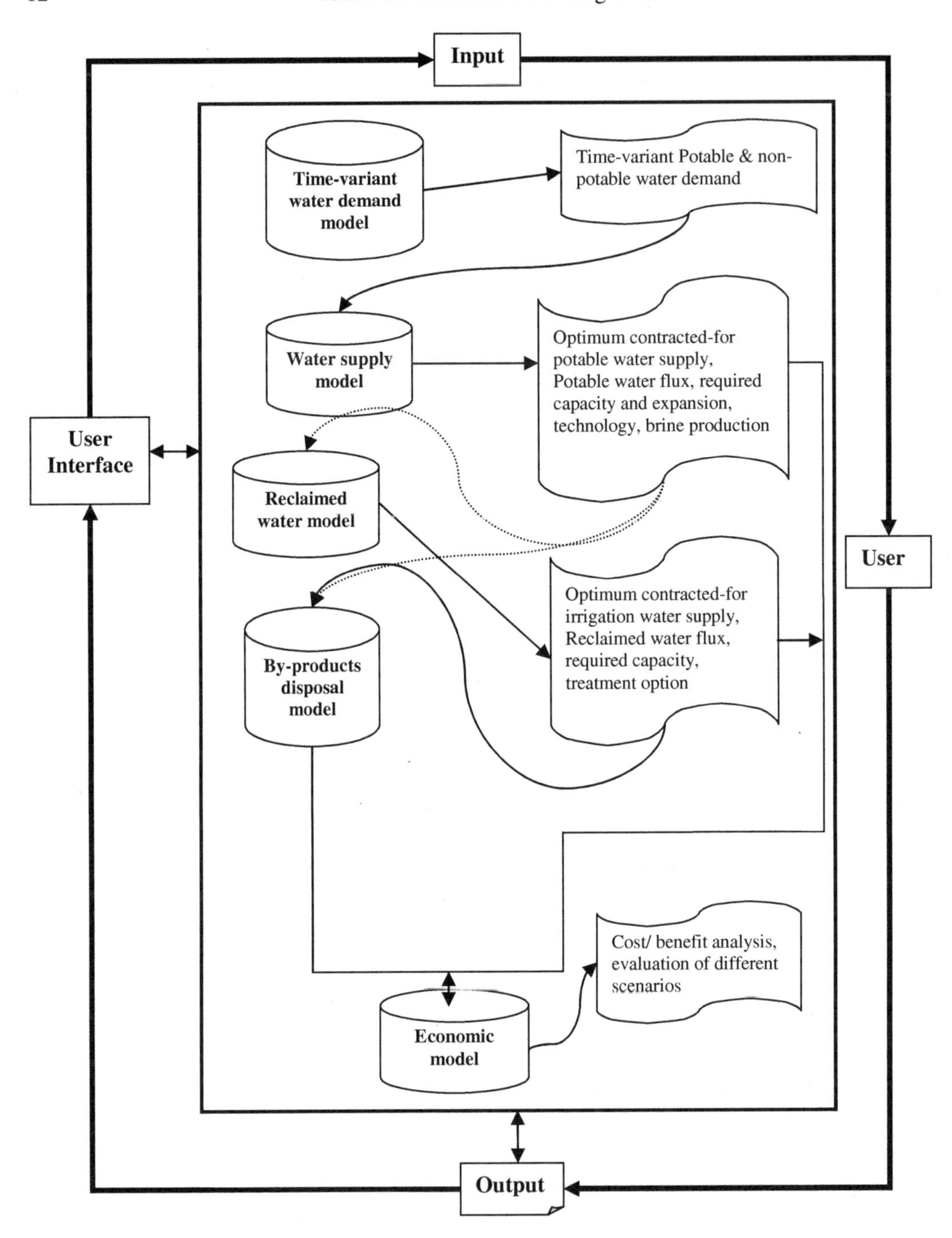

Figure 1.3 The proposed technical- economic modeling tool and the coupling of mathematical models

Innovation

The innovative points in the research are:

1. The developed technical-economic modeling tool is an integrated model including both desalination (as the main potable water resource) and reclaimed wastewater (as the main non-potable water resource). Previous studies have dealt with these sources as marginal alternatives and have not tackled the interrelation between them.

2. The technical-economic modeling tool will expand over previous models by including parameters and decision variables relevant for the feasible implementation of IWRM; for instance, service domain of project, optimum capacity and planned expansion, optimum contractual agreement for a hotel, selection of treatment technology, energy sources and economic costs of different brine disposal methods.

3. The technical-economic modeling tool will focus on tackling highly time-variant potable and non-potable water demand;

4. The technical-economic modeling tool offers a practical approach to the development of IWRM projects in arid coastal regions. In Egypt, this methodology has not been applied in that field before, and on that scale (project level). Available models do not approach the problem as a whole and do not suit local conditions and constraints.

Thesis Outline

The current chapter starts with discussing the importance of integrated water resources management and the need for a tool to aid decision making. This is followed by the identification of the research problems and defining the scope, objectives and relevance of the research. Subsequently, a literature review is presented comparing previous work done with proposed work in this research and highlighting the innovation of this research.

Chapter 2 presents information on the case study, i.e. the city of Sharm El Sheikh (Sharm) at the Red Sea in South Sinai, Egypt. Special attention is given to water demand, water supply options in the region, current water resources management practices, and shortfalls including increasing water shortages and price rises as well as environmental degradation which would impact the tourism industry. Strategies are outlined which could be undertaken to improve the situation (based on Lamei *et al.*, 2009a).

Chapter 3 presents a cost comparison for two options to supply water of drinking water quality: Option 1 (Desalination with the reverse osmosis technology), or Option 2 (Long-distance water piping from the Nile), for the case of Sharm. To analyze costs for Option 1, RO desalination plants costs (capital and O&M) for 14 RO plants in Egypt and 7 elsewhere for comparison were compiled. For Option 2, cost data for four long-distance piping projects in Egypt which pump groundwater or treated Nile water to cities in South Sinai including Sharm were presented (based on Lamei *et al.*, 2008a).

Chapter 4 presents unit production costs and energy costs for 21 RO desalination plants in the region. An equation is proposed to estimate the unit production costs of RO desalination plants as a function of plant capacity, price of energy and specific energy consumption. This equation is used to calculate unit production costs for desalinated water using photovoltaic (PV) solar energy based on current and future PV module prices. Unit production costs of desalination plants using solar energy are compared with conventionally generated electricity considering different prices for electricity. This chapter discusses at which electricity price solar energy can be considered economical to be used for RO desalination; this is independent of the RO plant capacity (based on Lamei *et al.*, 2008b)

Chapter 5 presents an analysis of the water demand by the tourism industry in arid coastal regions, i.e. hotels and related services. A model is developed to calculate time-variant water demand by the tourism industry. The model provides investors (public and private sectors) in water resources projects with the means to estimate current and future water demand required by existing or new tourism facilities. Also, it considers the impact of introducing demand management to reduce water consumption in hotel facilities (based on Lamei *et al.*, 2006)

Chapter 6 presents the setup of potable water supply to hotels in the case study region. Private water companies supply these hotels with their domestic water needs. There is normally a contractual agreement stating a minimum requirement that has to be supplied by the water company and that the hotel management has to pay for regardless of its actual consumption ("contracted-for water supply"). This paper describes an optimization model to determine what value a hotel should choose for its contracted-for potable water supply in order to minimize its total annual water costs as a function of occupancy rate and hotel characteristics (based on Lamei *et al.*, 2009b).

Chapter 7 presents a model that was developed using Excel macros to perform dynamic optimization with the objective function to maximize present value of total benefits over the lifetime of the RO desalination plant. The aim of the dynamic optimization is to solve for capacity expansion. The model can be used to test different scenarios to capture time-variant tourism demand and price uncertainties on investment decisions. With a RO desalination plant designed to satisfy only the contracted-for water supply, the water company would be missing out on potential benefits that could have been obtained selling water in periods of high demand. On the other hand, sizing the RO desalination plant to produce water to satisfy peak demand means incurring additional costs as well as having the plant partially idle during periods of average or low demand. Unit production cost of RO desalination plants varies according to the degree of operation of the plant. This fact has to be taken into consideration when calculating costs of RO desalination and when deciding on plant capacity in order to maximize net benefit (based on Lamei *et al.*, 2009c).

Chapter 8 describes a model to optimize the contracted-for irrigation water supply with the objective function to minimize total water cost to hotels. Contracting for 100% of the peak irrigation demand is the highest total cost option to a hotel management. Contracting for a portion of the peak irrigation demand and complementing the rest from desalination water is a cheaper option. A better option still is to complement the excess irrigation demand from the company that treats and sells wastewater, if available or from another wastewater treatment company at a higher cost (but at a cost cheaper than desalination water) mainly due to the high demand season and the additional cost of trucking. Having an on-site wastewater treatment plant presents further savings to the hotel. A sound strategy for water management would be to rely solely on treated wastewater on-site. This can be achieved by: increasing the efficiency of the irrigation system, reducing the area of high-water consuming plantation (e.g. turf grass) and/or shifting to drought resistant plants including less water-consuming or salt tolerant turf grass (based on Lamei *et al.*, submitted)

Chapter 9 discusses three commonly used disposal methods (sea disposal, well injection and evaporation ponds) with regard to their environmental impact and economic costs. The chemical characteristics of reject brine from five seawater RO plants in Egypt are presented with their possible impact on marine environment and groundwater aquifers. An overview of existing regulations governing brine disposal is given. Assessment criteria to evaluate the impact of the three different brine disposal methods, chemicals concentration limits, and best practices for brine management are suggested. Both direct and indirect costs of the three disposal methods are presented. Whereas the direct economic costs of brine disposal vary according to method and location (e.g. whether the plant is inland or coastal), the potential environmental costs are likely to be much higher. The selection of the most appropriate method thus depends on a correct estimate of the associated environmental costs and encourages (near) zero discharge and resource recovery (based on Lamei *et al.*, 2009d).

Chapter 10 concludes the thesis by outlining the extent to which the research problem was solved. A brief discussion of the theoretical and practical implications of the thesis research is provided. Finally, limitations and opportunities for further research are outlined.

2 Water resources management to satisfy high water demand in the arid Sharm El Sheikh, the Red Sea, Egypt

Abstract

Sharm El Sheikh (Sharm) in South Sinai, Egypt, is situated in an area of extreme aridity (annual rainfall between 20–50 mm/yr). It has been undergoing rapid development and attracts about one million tourists annually which results in an ever-increasing demand for water. The main source of water is desalinated seawater produced by two government-owned reverse osmosis (RO) plants, two centralised privately-owned RO plants and by about 50 decentralised small RO plants in hotels. The government-owned RO plants sell water to the local residents at a very low subsidized price while the two centralised private RO plants (owned by two different companies) charge commercial rates and raise prices considerably in the summer periods of high water demand. For all the plants, there are concerns over high energy consumption and the impact of brine discharge on the environment. Other sources of water in Sharm include tankers and pipes delivering groundwater from Al Tor (100 km distance) and treated domestic wastewater for landscape irrigation. The Egyptian Environmental Affairs Agency (EEAA) is not regulating and monitoring water management sufficiently. Increasing water shortages and price rises as well as environmental degradation would impact the tourism industry. This paper describes the current water resources management practices in Sharm, and outlines strategies which could be undertaken to improve the situation.

Keywords: Desalination; wastewater reuse; integrated water resources management; tourism; water demand; reverse osmosis.

This chapter is based on:
Lamei, A., van der Zaag, P. and von Münch, E. (2009a) Water resources management to satisfy high water demand in the arid Sharm El Sheikh on the Red Sea, *Desalination & Water treatment*, **1**, 299-306.

Introduction

Egypt has to cope with increased water demand due to population growth, rising standards of living, expansion of tourism, industrial output and agricultural activities. The South Sinai region in particular experiences water shortages. This area is important for Egypt's economic growth due to rapidly expanding tourism. A major tourist city in the arid environment of South Sinai is Sharm El Sheikh (Sharm), a popular Red Sea resort.

The purpose of this paper is to describe how Sharm's water demand is met and to highlight existing shortcomings of a haphazard approach to water resources management. Sharm can be seen as a typical example of a touristic city in an arid environment, with an urgent need for effective water resources management. Sharm is situated on the southern tip of the Sinai Peninsula with the Red Sea on one side and the mountains of Mount Sinai on the other. Sharm, "the jewel of Sinai", has year-round sunshine and popular beaches.

About one million tourists from Egypt and abroad visit Sharm each year. The current permanent population of Sharm was estimated to be 25,000 in 2006 but may be much higher due to non-registered temporary labourers. The annual growth rate for the local population is about 3.8% per annum. In May 2006 there were 65 hotels, mostly 3–5 star category, and 63 more hotels under construction within the city limits. Future construction will be outside the city limits (Khaled, 2008).

Little attempt has been made to minimize the water consumption of hotels even though the city is located in a region of extreme aridity (annual rainfall between 20–50 mm/yr) (Abou Rayan *et al.*, 2001), and has no groundwater resources. To meet rising demand, privately-owned seawater reverse osmosis (RO) plants are being built.

It is estimated that approximately 91% of the current average water demand in Sharm is from the tourism industry (hotels, restaurants, bars, shops, staff housing and landscape irrigation). The remaining 9% of the water demand is from the local population (Khaled, 2008). The per capita water consumption of the permanent residents is about 100-150 l/cap/d. However, the actual consumption is much higher reaching 250 l/cap/d, mainly due to wastage, absence of water meters, cheap subsidized water for residents and leakage from the distribution system which can reach up to 40% (Abd Al Latif, 2008).

The average tourist uses huge amounts of water ranging from 300 to 850 liters per day depending on hotel facilities and services, occupancy rates, ambient temperature, staff housing and irrigation area (Chapter 5).

Most of the data presented in this paper was obtained from the Sinai Development Authority (government) which is responsible for all development activities in South Sinai including water supply, wastewater, road building, etc. Data was also obtained from interviews with the Chief Engineers of several government-owned and private RO desalination and wastewater treatment plants in Sharm, and with hotel managers from eight 5-star hotels.

Water supply methods used in Sharm

There is a mixture of water supply methods used in Sharm (Figure 2.1). The city depends mostly on RO desalination (86% of the water supply) with the remainder being supplied by groundwater transported by tankers or long-distance pipelines.

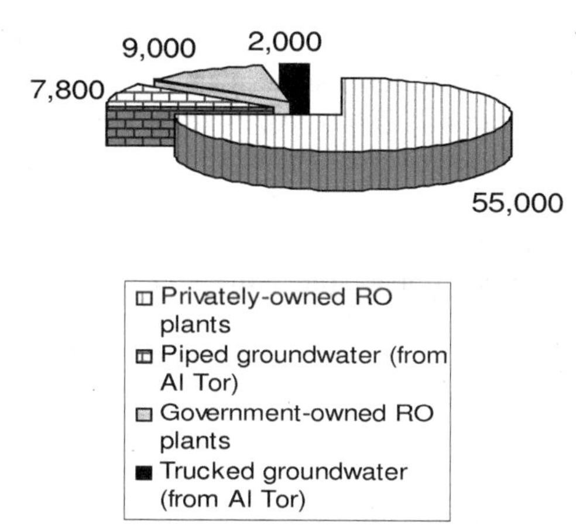

Figure 2.1 Average potable water production in Sharm for 2008 from different resources (in m^3/d). Total amount is 73,800 m^3/d (Khaled, 2008) . Treated wastewater reuse is not included.

Pipelines and tankers

In the 1970s, Sharm depended on a 100 km long water pipeline (diameter of 250 mm) transporting groundwater from the Al Tor 100 m thick fresh water aquifer (MWRI, 2005). Due to developments in Sharm requiring more water, tankers were added. With a further increase in local population and tourism, tankers were no longer a practical solution.

There are two other pipelines which bring treated Nile water to Sinai. At first these were intended to bring water also to Sharm, but all the water is now delivered to other locations and does not reach Sharm. The groundwater abstraction rates in Al Tor are now unsustainably high (in total 9,800 m^3/d of groundwater is pumped/trucked from Al Tor to Sharm). Wells are drying up due to the groundwater table going down as recharge is low in this region.

Reverse osmosis seawater desalination plants

RO desalination is organized by three types of plants (Table 2.1):

- Two government-owned RO plants;
- Two centralised private RO plants; and
- About 50 private RO plants in individual hotels.

The government-owned RO plants provide water to the local population and public buildings, but do not meet the full demand. Each residential district receives water for 2–4 h a day.

These plants have direct surface intakes for seawater. Their source water is of a lower quality than intake water from beach wells since beach wells can act as natural filters for seawater. This increases the production cost due to the extra pre-treatment needed.

The two centralised private RO plants provide water to those hotels which do not have their own desalination plants, or where the water demand exceeds the capacity of their own RO plants.

Many hotels (about 50) have their *own private desalination plants* to achieve autonomy. These decentralised RO plants are not monitored and regulated. Intake water is sourced from beach wells (30 m depth) which in many cases are not very far apart (less than 50 m) from brine disposal wells (60 m depth) regardless of the suitability of the local geological formation, thus raising the salinity of the intake water (from 45,000 to 52,000 ppm). Increased salinity of intake water results in increased energy requirements of the RO plant (Hafez and El Manharawy, 2002). Apart from well disposal, brine is disposed into the sea which is likely to have adverse impact on sensitive marine life. Brine contains chemicals concentrated during the pre-treatment, and brine disposal was shown to cause damage and reef degradation in the area of Ras Mohamed National Park in Sharm (UNEP/PERSGA, 1997). Similar impact on fauna and flora has been observed in the vicinity of the brine outlet at an RO plant in Ashkelon, Israel (Einav and Lokiec, 2003). In 1994 the Egyptian Ministry of State for Environmental Affairs (EEAA) set up Law 4 (MSEA, 2006) forbidding brine disposal into the sea but this law is not yet well enforced.

Table 2.1 RO desalination plants in Sharm

Type of plant	Intake source	Capacity (m^3/d)	Customers	Unit production cost (US$/$m^3$)	Selling Price (US$/$m^3$)	Comments
2 government-owned RO plants	surface intake		Local residents & public buildings	1.2[a]	0.05-1.21 (see Table 2.4)	Production is not enough to satisfy local population, therefore does not sell to hotels
Al Ta'meer		6,000				
Al Shabab		3,000				
2 private centralised RO plants	beach wells		Hotels	0.9[b]	1.6-2.5 (the higher value is charged during peak summer months)	Licensed to produce and sell water to others
South Sinai Water Co.		17,000				
Ridgewood		7,000				
50 private decentralised RO plants	(varied)	31,000 (typical value about 600 m^3/d per plant)	Hotels	1.2-2.9[c] (based on 14 plants)	N/A	Located within individual hotels for own usage
Total 54		64,000				

[a] (Abd Al Latif, 2008)
[b] (Girgis, 2008)
[c] (Chapter 3)

Treated effluent reuse

Domestic wastewater is treated with the following two aims:

- to produce water of suitable quality for hotel landscape irrigation (apart from treated effluent, RO product water is also being used for landscape irrigation in many hotels because of low quality of treated effluent or insufficient amount of treated effluent); and/or

- to allow discharge to the environment according to the Egyptian regulations.

Domestic wastewater is treated in wastewater treatment plants (WWTPs) in Sharm (Table 2.2)

- One government-owned wastewater treatment plant;
- One private centralised wastewater treatment plant;
- Fifty private decentralised wastewater treatment plants in individual hotels.

Table 2.2 Wastewater treatment plants in Sharm

Type of plant	Capacity (m^3/d)	Type of plant	Use of treated effluent
1 government-owned WWTP	15,000	Waste stabilization ponds	Half the flow is treated and used to irrigate 40 ha of government-owned forest plantation (average raw sewage inflow in 2006: 8,500 m^3/d). Rest of flow is diverted to the private centralised WWTP
1 private centralised WWTP (South Sinai Water Co.)	6000	Aerobic/anaerobic treatment	Treated effluent used to treat a golf course (100 ha) owned by South Sinai Water Co.
50 private decentralised WWTPs [a]	400 per plant	Aerobic/anaerobic treatment	Treated effluent used to irrigate hotel landscape
Total	41,000		

[a] 80% of total hotels in Sharm have their own WWTP; hence there are about 50 small WWTPs (Khaled, 2008)

The interviewed hotel managers claim that the effluent quality from their WWTPs complies with Category 1 of the Egyptian Decree No. 44 of the year 2000 (which is an amendment to Law 4), which regulates wastewater reuse for irrigation (hotels do their own effluent quality monitoring, and the EEAA is supposed to perform regular checks on the wastewater treatment plants of the hotels but is failing to do so). The hotel managers were not willing or able to disclose effluent quality data from their wastewater treatment plants.

Agriculture activities are classified into three categories (Table 2.3): Category 1: landscape irrigation; Category 2: animal feed plantation, dried seeds, flowers, fruits with skin (commercially processed) e.g. lemon, dates; Category 3: woods. The Decree forbids the usage of treated wastewater for irrigation of edible vegetables (whether eaten raw or cooked), as well as fruits eaten raw without a skin, e.g. grapes. According to the Decree, Category 1 requires the highest level of treatment: secondary treatment, sand filtration and disinfection (tertiary treatment). Category 2 requires only secondary treatment, e.g. activated sludge, oxidation ditches, trickling filters and stabilization ponds. Category 3 requires only primary treatment.

USEPA guidelines for open landscape irrigation are equivalent to Category 1 in the Egyptian standards, while USEPA guidelines for restricted landscape irrigation are equivalent to Category 2. The Egyptian standards for treated effluent for reuse in landscape irrigation (Category 1) are less stringent than suggested guidelines from

USEPA (open access) for *E. coli*. While USEPA requires no *E. coli* in treated wastewater, the Egyptian standards allow up to 1000/100 ml. The BOD limit in the Egyptian standards (Category 1) is also higher than the USEPA guidelines (open access).

The treated effluent sample from the government-owned WWTP (used for woods irrigation) adhered to Egyptian regulations (Category 3) but not to USEPA guidelines (either open or restricted access) in case of BOD. There was no test performed for *E. coli*.

Law 4 for year 1994 mandates that excess sludge from WWTPs is disposed in sanitary landfills. The centralised government-owned WWTP is relatively new (built in 2002) and is currently operating at only half its capacity. Desludging of the ponds will only be required in several years. In the case of small decentralised WWTPs, excess sludge is commonly disposed of in the backyards of hotels causing health and environmental hazards. This is against the EEAA's environmental regulations but these regulations are not sufficiently enforced in Sharm. The monitoring is easier for the centralised government-owned WWTP than for the many small WWTPs.

Table 2.3 Egyptian standards for the three categories of agriculture activities along with suggested guidelines from USEPA for open and restricted access landscape irrigation and a sample from treated effluent from the government-owned WWTP in Sharm

Parameter	Egyptian Standards			USEPA Guidelines [a]		Effluent quality from government-owned WWTP (One sample)
	Category 1	Category 2	Category 3	Open landscape irrigation	Restricted access landscape irrigation	
pH	-	-	6-10	6-9	6-9	-
BOD (mg/l)	≤20	≤60	≤400	≤10	≤30	105
E.coli (no./100 ml)	1000	5000	-	none	200	-
TSS (mg/l)	20	50	250	-	30	26
Residual chlorine (mg/l)	-	-	-	≥1	≥1	340
COD (mg/l)	40	80	≤700	-	-	222
Oil and grease (mg/l)	≤5	≤10	≤100	-	-	2.2
TDS (mg/l)	2000	2000	2500	-	-	138
Temperature (°C)	-	-	≤40	-	-	-

[a] (Tchobanoglous *et al.*, 2003)

Water supply costs in Sharm

Potable water from pipelines, tankers and RO desalination plants

In this extremely arid region, pumping water from the Nile is using up precious Nile water on one hand. On the other hand, the pipes' capital and O&M costs are substantial compared to RO desalination costs. The unit production cost of long-distance piped water for a distance of 368 km (distance from nearest Nile water source) and a capacity of 15,000 m^3/d, for instance, is about 1.9 US\$/m^3 compared to an international standard trend of less than 1US\$/m^3 for RO desalinated water (Chapter 3). The price of tankered groundwater is approximately 2.6 US\$/m^3 which is higher than RO product water (Abd Al Latif, 2008; Mohsen, 2007).

The government-owned RO desalination plants sell water at a much lower rate than the actual cost (see Table 2.1). Table 2.4 shows the government's tariff system for RO product water for different clients. The actual unit production cost of product water for these RO plants is approximately 1.2 US$/m^3 whereas the selling price is as low as 0.05 US$/m^3.

In the Canary Islands in Spain, for instance, the government also offers state support to desalination plants to obtain water prices equal to that of water used by households in all of Spain. This subvention has lead to a higher level of water production with fewer interruptions for the users. However, the subvention is inversely proportional to the energy consumption of the plant, and proportional to the efficiency of the water supply network in order to encourage optimised plant performance (Gasco, 2004). A similar system could be devised as a subsidy scheme for the government-owned RO plants in Sharm.

Table 2.4 Tariff system for RO product water from government-owned RO plants

Usage	Consumption (m^3/month)	Prices (US$/m^3)
Domestic (public)	1-30	0.05
Domestic (public)	30-50	0.06
Domestic (public)	50 and above	0.19
Construction	No limit	0.19
Airport	No limit	0.19
Hotels/Domestic (private)	No limit	1.21

The unit production cost of the government-owned RO plants is high (though lower than those of the private decentralised RO plants) due to excessive use of chemicals and additional pre-treatment stemming from the surface intake as opposed to the beach wells; also their capacities are smaller than the private centralised RO plants.

The selling price of RO product water from the private centralised plants is relatively high (1.6-2.5 US$/m^3 as per Table 2.1; the higher value is charged during the peak summer months) and hotels are forced to buy to satisfy their demand unless they have enough water from their own RO plants. The lowest unit production cost is achieved by the private centralised RO plants due to their more continuous and optimal operation. These plants take their intake water from beach wells therefore reducing pre-treatment cost.

The current unit production costs of the private decentralised RO desalination plants in Sharm are high compared to international trends (up to 2.9 US$/m^3 in Sharm versus less than 1 US$/m^3 worldwide) (Chapter 3). Reasons for elevated costs of RO desalinated water in Egypt are: (a) old plants; (b) technical problems due to lack of experience and poor maintenance, (c) small-capacity desalination plants (no economies of scale) and (d) large variations in operating time. Especially the latter appears to be a significant factor as water production cost is inversely proportional to operating time (Voivontas *et al.*, 2003).

Due to variations in hotel occupancy rates in Sharm, water demand varies throughout the year. As a result, the small decentralised RO plants only operate at full capacity during relatively short periods of time, and are frequently ramped down. This interrupted service affects the quality of water and the lifetime of membranes and accordingly increases their costs. This problem is faced by tourism-dominated arid regions around the world.

Non-potable water from WWTPs

The government charges the local population for the wastewater collection and treatment by charging an additional 33% of the potable water bill. If the government was selling potable water to hotels, it would have charged them a wastewater discharge fee of 0.4 US$/m^3 of potable water used. However, as the government-owned RO plants do not provide potable water to hotels, the government does not charge them for wastewater collection and treatment although nearly all hotels are in fact connected to the public sewer network (though not all of them discharge to the network).

Table 2.5 shows unit capital costs for five wastewater treatment plants (all waste stabilization ponds) in South Sinai. The O&M cost is about 0.2 US$/m^3 including maintenance and spare parts, labour and chemicals (chlorine). The unit production cost of treated wastewater (for the government-owned WWTP in Sharm) is calculated to be about 0.25 US$/m^3 based on 30 years lifetime and 8% interest rate.

Since this cost is much lower than the unit cost of RO product water, it of course makes economic sense for the hotels to use treated wastewater for landscape irrigation instead of RO product water.

Table 2.5 Capacity and capital costs of waste stabilization ponds in South Sinai, Egypt-all owned by the Sinai Development Authority (costs are adjusted for year 2008[a])

No.	Year built	Location	Capacity (m^3/d)	Capital cost (Million US$)	Unit capital cost (US$/m^3/d)
1	2002	Sharm El Sheikh	15,000	2.69	179
2	2001	Ras Sudr	5,000	1.52	303
3	2001	Abou Zeinema	5,000	1.29	258
4	2001	Abu Rudeis	5,000	1.26	252
5	2001	Saint Catherine	1,000	0.54	542

[a] Calculated based on the United States Consumer Price Index

Options for improved integrated water resources management

As shown above, the current water resources management in the city of Sharm is facing significant challenges with respect to institutional, economic, technical and environmental aspects. These challenges, and possible solutions, are summarised in Table 2.6. Optimised management of water resources in the area would lead to better services and possibly lower prices (for RO product water sold to hotels), positively impacting the tourism sector (lower hotel running costs), as well as the communities that directly depend on this industry.

A technical-economic modeling tool is currently being developed to aid decision makers, including public and private investors, in the design and assessment (both economically and environmentally) of future water supply infrastructure projects for arid coastal regions. For example, the modeling tool could predict the optimal capacity and required expansion of RO plants based on input values provided by the user, such as time-variant water demand and hotel characteristics.

Conclusions

The city of Sharm is a typical example of a tourisitc city located in an extremely arid environment with a haphazard approach to water resources management. Water supply to Sharm is mainly from RO desalination complemented by groundwater transported from Al Tor by tankers or long-distance pipelines and by treated domestic wastewater (for landscape irrigation).

Desalinated water is provided by two government-owned RO plants, two centralised privately-owned RO plants and by about 50 decentralised small RO plants. The government-owned RO plants sell water at a very low subsidized price to the local residents while the two centralised private RO plants (owned by two different companies) could control the market and raise prices considerably especially in periods of high water demand (selling price ranges from 1.6 to 2.5 US$/m^3). All of these RO plants cause environmental problems due to high energy consumption per m^3 of water produced and the impact of uncontrolled brine disposal on the environment.

The EEAA does not regulate and monitor water management in Sharm sufficiently. In the longer term, this situation is likely to lead to further water shortages and price rises as well as environmental degradation which would impact the tourism industry.

Measures including stronger enforcement of related laws; evaluating the impact of erection of (semi-) centralised desalination and wastewater treatment plants; and considering alternative management methods for brine and sludge can help in achieving a more sustainable water resources management.

Acknowledgements
We would like to thank Prof. Magdy Abou Rayan, Mansoura University, Egypt and Dr. Ibrahim Khaled, Sinai Development Authority, Egypt for kindly providing valuable data for this research.

Table 2.6 Summary of challenges and suggested actions related to existing water resources management in Sharm

Category	Problems (in bold: most important problems)	Suggested actions for improvements (in bold: most important actions)
Institutional	• **Brine disposal at private decentralised desalination plants is not effectively regulated and monitored** • **Lack of effective monitoring for sludge disposal in private decentralised wastewater treatment plants** • Local regulations are less stringent than international guidelines (USEPA) for wastewater treatment (e.g. *E.coli*) for Categories 1 and 2 • Egyptian law does not allow irrigation of edible vegetables (and skinless fruits) with treated wastewater	• **Stronger enforcement of law 4 for year 1994 by the EEAA and introducing of fines for violation of the law** • Review of Egyptian standards for reuse of treated wastewater to further improve quality of treated effluent to minimize health and environmental risks
Economic	• Cheap subsidized water for local residents • Hotels are not charged for being connected to the public sewers • Insufficient cost recovery from government-owned plants (desalination and wastewater treatment) • **High unit production costs for private decentralised RO desalination plants** • **The two private water supply companies may dominate the market**	• Adjust subsidy system for local residents to provide incentives to stop wasting water; install water meters • Better billing system for wastewater discharge to public sewers (by hotels) • **Evaluate if erection of more (semi-) centralised plants (desalination and wastewater treatment) to replace the small decentralised plants would reduce production and treatment costs due to economies of scale (using newly developed DSS) and allowing for better monitoring and enforcement of regulations**
Technical	• Interrupted water supply services for local residents • Water consumption by residents is high and not metered • High leakage in public distribution network • Increased salinity of desalination plants beach wells due to close proximity to brine disposal wells • Aquifer underlying Al Tor is being exhausted (groundwater table is going down) • Hotels are irrigating with RO product water when own treated effluent has low quality or insufficient amount	• Put a leakage detection programme in place, repair leaks in water supply network • Construction of deep brine disposal wells to avoid mixing with source water from shallow intake wells (and proper investigation of subsoil formation to avoid interference between supply and reject water) • Roll out a programme of water demand management for hotels, for example low water use appliances; low-flush toilets; use drought-resistant desert plants for landscaping to reduce on irrigation water use
Environmental	• **High energy requirements per cubic metre of water produced** • **Effluent quality from decentralised wastewater treatment plants is not monitored sufficiently** • **Negative impact on seawater quality from brine disposal in the Red Sea is likely** • Negative environmental impact due to open disposal of sludge in hotel backyards is likely	• **Evaluate alternative methods for brine and sludge management to reduce environmental impacts (using newly developed DSS)**

3 Basic cost equations to estimate unit production costs for RO desalination and long-distance piping to supply water to tourism-dominated arid coastal regions of Egypt

Abstract

An arid climate with limited water resources and a growing tourism industry lead to water shortages in many coastal zones. Due to increasing demand, alternatives have to be found, e.g. desalination and long-distance water piping (long-distance defined here as equal to or further than 30 km), ecological sanitation, wastewater reuse or water demand management. This paper presents a cost comparison for two options to supply water of drinking water quality: Option 1 (Desalination with the reverse osmosis technology), or Option 2 (Long-distance water piping from the Nile), for the case of the touristic city of Sharm El Sheikh (Sharm) at the Red Sea in South Sinai, Egypt.

Available water resources and current as well as future water demand figures for Sharm are presented. 91% of the current water demand stems from tourism; water is supplied mainly by privately owned RO desalination plants (86%). To analyze costs for Option 1, we compiled RO desalination plants costs (capital and O&M) for 14 RO plants in Egypt and 7 elsewhere for comparison.

Unit production cost (US$/m^3) of water from small RO desalination plants in Egypt is in most cases lower than international trends for similar small capacity plants (250 to 5,000 m^3/d), but unit O&M costs are higher. For Option 2, we present cost data for four long-distance piping projects in Egypt which pump groundwater or treated Nile water to cities in South Sinai including Sharm. We found that unit capital costs for those pipelines which are longer than 140 km are in fact above the cost of a possible RO desalination plant at any flow capacity. For unit production cost, desalination costs are lower than long-distance piping starting from pipelines with 300 km length or more and capacity $\geq$ 2000 m^3/d. Empirical basic cost equations are produced to calculate unit capital cost (US$/m^3/d) and unit production cost (US$/m^3) for both options in dependence of capacity for Option 1, and capacity and pipe length for Option 2.

Keywords: Water demand; capital and O&M costs; Sharm El Sheikh; Nile water.

This chapter is based on:
Lamei, A. van der Zaag, P. and von Münch, E. (2008a) Basic cost equations to estimate unit production costs for RO desalination and long-distance piping to supply water to tourism-dominated arid coastal regions of Egypt. *Desalination*, **225,** 1-12.

Introduction

The Middle East and North Africa regions not only have the world's lowest per capita availability of water resources but also the highest rate of reduction in these resources (El Fadel and Alameddine, 2005). Water scarcity is particularly increasing in those coastal zones which are characterized by an arid climate (less than 100 mm rainfall per year), and a thriving tourism industry with a high water demand throughout the year. Egypt (with a current population of 78 million) is among those countries which face serious water scarcity (CIA, 2005). It is located furthest downstream in the Nile basin with an extremely arid climate. The annual share of Egypt from Nile water is 56 billion m^3. The average per capita share is 711 m^3/cap/a (Elarabawy *et al.*, 2000).

In South Sinai, which is on the Red Sea in Egypt (see Figure 3.1), tourism is the dominating industry. In this region, rainfall is extremely low; Sharm El Sheikh (Sharm) has an annual rainfall between 20–50 mm/yr (Abou Rayan *et al.*, 2001).

This region is facing continuous economic growth which results in increasing water consumption. The shortfall between water demand and water availability is expected to reach 365 Mm^3/y in South Sinai and along the Red Sea coast by the year 2020 (Hafez and El Manharawy, 2002). Since conventional surface and groundwater resources are limited, water can be sourced by either RO desalination of sea or brackish water, or by long-distance tanker trucks or a pipeline (long-distance defined here as equal to or further than 30 km). Other options to address the expected water shortfall could be an ecological sanitation approach including water demand management and wastewater reuse. RO desalination is increasingly being used to address this shortfall in this region.

The scope of this paper is limited to the analysis of two options; Option 1: RO desalination and Option 2: long-distance piping ($\geq$30 km). These two options are currently the most important ones in the city of Sharm.

For both options, we developed basic cost equations to be used by decision-makers, and we compared the costs in Egypt to world-wide trends for Option 1. All cost figures quoted in this paper are in US$ of 2001 (unless otherwise indicated) calculated based on the United States Consumer Price Index:
(http://minneapolisfed.org/Research/data/us/calc/index.cfm#calc).

The base year 2001 was used here because this was the year used in (GWI, 2006) for the extensive data set on costs of RO desalination plants.

Water demand in Sharm

Data on water resources and demand in the city of Sharm presented here were obtained from the Sinai Development Authority which is concerned with the development of the Sinai region, in terms of provision of clean water, wastewater treatment and road construction. The data in this section was taken from (Khaled, 2008) unless otherwise indicated.

Sharm is a prosperous Red Sea resort remote from the Nile. There were (in May 2006) 65 hotels in Sharm mostly falling in the 3 to 5 star category, and 63 more hotels under construction within the city limits. Future construction will be outside the city limits. Each year, over one million tourists from Egypt and abroad visit Sharm. The current permanent population of Sharm is about 25,000. The growth rate for the local population is about 3.8% per annum.

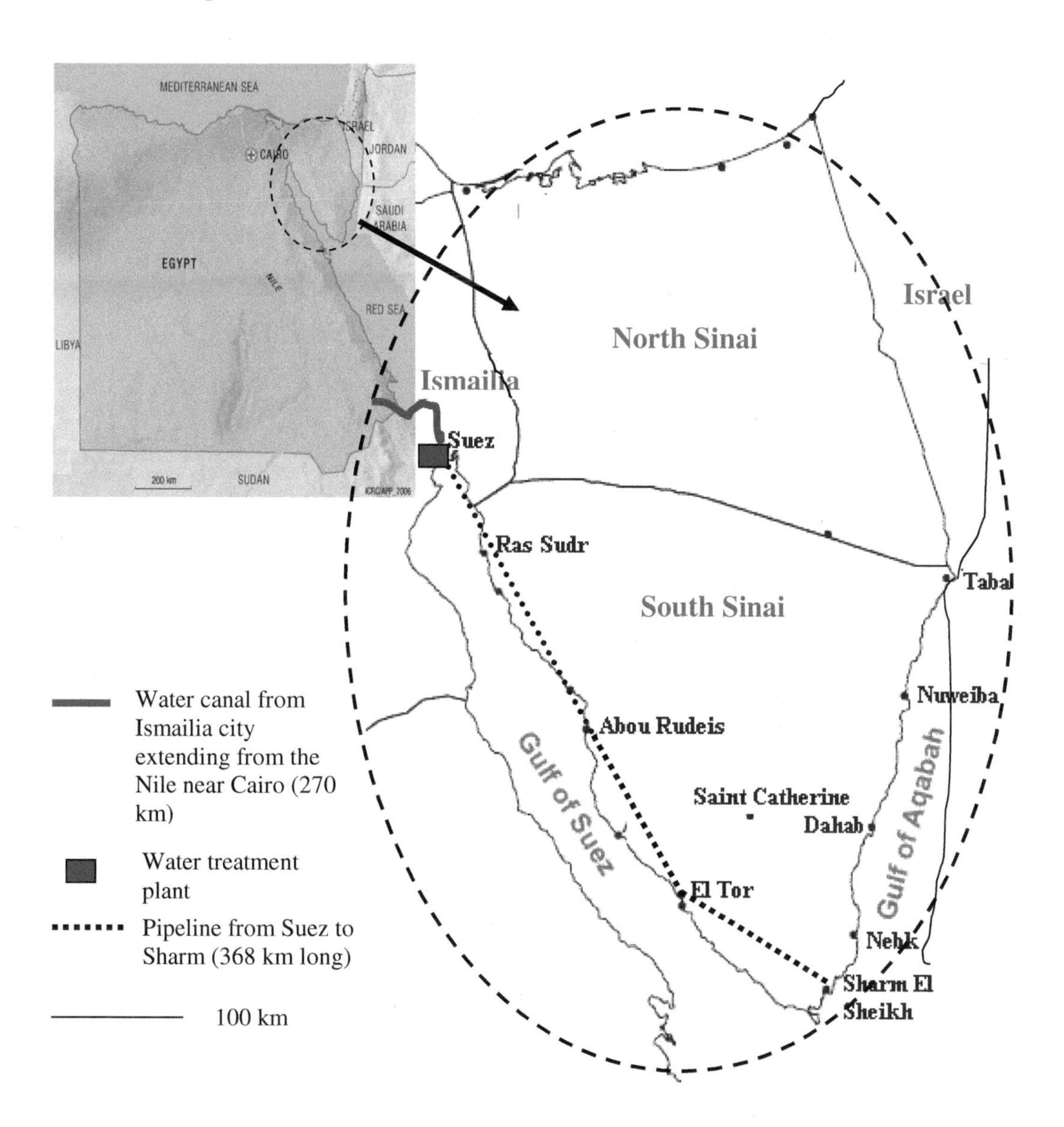

Figure 3.1 Map of Egypt showing location of Sharm and major water pipeline (http://www.icrc.org/Web/eng)

Figure 3.2 shows the water production in Sharm in 2006 from different sources with a total of 69,300 m^3/d (86 % of this is obtained via RO desalination plants). There are also two long-distance water pipelines, one to transport treated groundwater from El Tor (100 km away) and the other to transport treated Nile water from Suez (368 km away) to Sharm. However, at present, the latter pipe is used instead to transport groundwater from El Tor city (no Nile water is currently being transported to Sharm). A pipeline is currently under construction to double the capacity of the pipeline conveying Nile water from Suez to Sharm (Table 3.3). A small amount of groundwater is also transported from El Tor to Sharm by tanker trucks.

91 % of the current average water demand in Sharm is from the tourism industry (including hotels, restaurants, bars, shops, staff housing and landscape irrigation). The remaining 9% of the water demand is from the domestic population. The total number of tourist-related buildings is approximately 200.

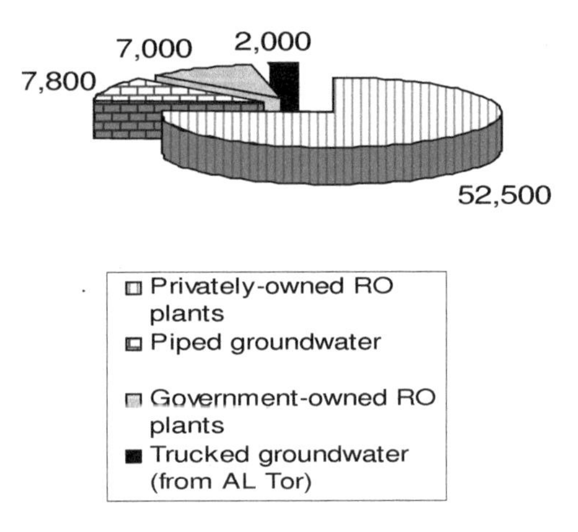

Figure 3.2 Average water production in Sharm for 2006 from different resources (in m^3/d)

Water demand in tourism-dominated coastal regions is proportional to hotel room occupancy rates which vary according to time of the year and marketing efforts of each hotel. The conventional tourist uses huge amounts of water so that the equivalent tourist water consumption ranges from 300 to 850 liter per day per occupied bed depending on individual hotel facilities and services, occupancy rates, temperature, staff housing and irrigated area (Chapter 5).

The current annual increase in water demand due to construction of new hotels in Sharm is 11% per annum. This is expected to slow down in the near future to about 5% per annum. Shortages are therefore estimated to reach 3 Mm^3/y for Sharm by the year 2010 (Abou Rayan *et al.*, 2001).

To meet this rapidly growing water demand, city planners are now considering two main options (RO desalination or long-distance piping) or a combination of these. Basic cost equations for these two options are presented below. Other innovative options are outside the scope of this paper as explained earlier.

Option 1: Reverse osmosis desalination (Basic cost equations)

RO desalination used to be considered a very expensive technology. However, recent advances in RO desalination technology and larger capacity plants are reducing the production costs per cubic meter.

Hafez and El Manharawy (2002) compiled RO desalination costs from five plants (capacity ranging from 250 to 4,800 m^3/d) in South Sinai in Egypt and compared them to costs from seven other plants (capacity ranging from 7,000 to 50,000 m^3/d) in Saudi Arabia, Libya, Tunisia and Cyprus. The authors reached the conclusion that unit capital and unit O&M costs of RO desalination plants in Egypt are more expensive than worldwide trends. We do not agree with this finding as will be shown below.

Another more comprehensive source of desalination plant costs is given in (GWI, 2006). The report assumes that unit production costs (in US$/m^3) can be split up into 40% for annual cost of capital per unit and 60% for unit O&M costs.

We used data from (Hafez and El Manharawy, 2002) together with our own data and available literature to compare RO costs in Egypt with international costs.

Table 3.1 shows a compilation of costs for 14 desalination plants in Egypt (13 of which are in South Sinai region), and seven other plants outside of Egypt in the Mediterranean region and Saudi Arabia. The breakdown of total investment cost into RO plant cost (i.e. intake system, pre-treatment, RO desalination, post-treatment, brine disposal and engineering) and infrastructure cost (cost of land, building and site work) is provided for those plants where this information was available.

The annuity factor, a, was determined from the following equation:

$$a = \frac{i \times (1+i)^n}{(1+i)^n - 1} \qquad \text{Equation 3.1}$$

where i is the discount rate (we used 8 %) and n is the economic plant life (we used 10 years lifetime for the RO plant equipment and 20 years lifetime for infrastructure).

The annual cost of capital can then be calculated by multiplying the investment cost with the annuity factor, a. To convert this to a unit cost, it is further divided by the annual output of the plant (assumed 90% of capacity).

Operation and maintenance (O&M) costs include pre- and post-treatment chemicals, membrane replacement, energy, labour, brine disposal and administration (a detailed breakdown for some plants is provided in Table 3.2). Unit production cost (US$/m^3) is a summation of the annual cost of capital (US$/yr) divided by the annual output of the plant (m^3/yr), plus annual O&M costs (US$/yr) divided by the annual plant output (m^3/yr).

From Table 3.1 it can be seen that unit production costs are high compared to what is currently the industry standard (about 0.5 to 1 US$/m^3). This can be attributed to the fact that most of these plants are quite old (they were built in the 1990's) and that they are small in size. Regarding O&M, actual values of O&M are higher than the estimated O&M costs except for plants number 2, 11 and 15. It appears that for smaller plants (< 5,000 m^3/d) the estimation of unit O&M costs via the 60% of investment cost value under estimates the actual unit O&M costs. Another explanation could be that, for reasons described in Table 3.2 below, the unit O&M costs in Egypt are simply higher than international trends.

Figure 3.3 shows a comparison between unit production costs of RO plants as listed in (GWI, 2006) with costs of plants presented in this chapter. This figure shows that the unit production cost of RO desalination in Egypt is comparable or even lower than international plants (using estimated or actual O&M figures, respectively). This is contrary to the findings in (Hafez and El Manharawy, 2002) that unit production costs of RO desalination plants in Egypt are higher than worldwide. Those authors did not compare the RO plants in Egypt (250 to 5,000 m^3/d) with similar plant capacities but with larger ones (7,000 m^3/d to 50,000 m^3/d).

Table 3.1 RO desalination costs in Egypt and in the Mediterranean region and Saudi Arabia (all in 2001 US$) (Abou Rayan *et al.*, 2001; Abou Rayan *et al.*, 2003; Abou Zeid, 2006; Hafez and El Manharawy, 2002; Khaled, 2008)

No	Capacity (m^3/d) (Q_w)	Location	Year of commissioning	Investment (million US$)			Unit capital costs[a] (US$/$m^3$/d) ($C_c$)	Annual cost of capital per unit[b] (US$/$m^3$)	Estimated unit O&M costs[c] (US$/$m^3$)	Actual unit O&M costs (US$/$m^3$)	Unit production costs with estimated O&M (US$/$m^3$) ($C_p$)	Unit production costs with actual O&M (US$/$m^3$) ($C_p$)
				RO	Infra-structure	Total						
	C1	C2	C3	C4	C5	C6=C4+C5	C7=C6/C1	C8	C9=C8*6/4	C10	C11=C8+C9	C12=C8+C10
1	250	Hurgada (Egypt)	N/A	0.40	0.01	0.40	1659	0.74	1.12	2.47	1.86	3.21
2	300	Nuweiba (Egypt)	N/A	N/A	N/A	0.59	1961	0.89	1.33	0.93	2.22	1.82
3	350	Red Sea (Egypt)	2004	N/A	N/A	0.33	950	0.43	0.65	0.93	1.08	1.36
4	500	Safaga (Egypt)	N/A	0.77	0.03	0.77	1605	0.72	1.08	2.22	1.80	2.94
5	500	Hurghada (Egypt)	1997	N/A	N/A	0.54	1074	0.49	0.73	0.93	1.22	1.42
6	500	Matrouh (Egypt)	1998	0.33	0.14	0.33	941	0.39	0.58	0.87	0.97	1.25
7	500	Dahab (Egypt)	1995	N/A	N/A	N/A	N/A	N/A	N/A	N/A	N/A	2.57
8	600	Taba (Egypt)	1986	N/A	N/A	N/A	N/A	N/A	N/A	N/A	N/A	2.95
9	2,000	El Tor (Egypt)	N/A	2.41	0.22	2.41	1313	0.58	0.87	1.65	1.45	2.23
10	3,500	Sharm (Egypt)	N/A	4.27	0.25	4.27	1291	0.58	0.86	1.51	1.44	2.08
11	4,000	Gulf of Aqabah (Egypt)	1995	N/A	N/A	6.46	1614	0.73	1.10	0.93	1.83	1.67
12	4,000	Sharm (Egypt)	1998	N/A	N/A	7.03	1756	0.80	1.20	1.23	1.99	2.02
13	4,800	Hurgada (Egypt)	N/A	4.23	0.78	4.23	1044	0.45	0.68	1.10	1.13	1.55
14	5,000	Hurgada (Egypt)	1997	5.35	1.45	6.80	1360	0.58	0.86	0.97	1.44	1.54
15	7,000	Libya	N/A	N/A	N/A	7.92	1131	0.51	0.77	0.70	1.28	1.21
16	10,000	Tunis	N/A	N/A	N/A	9.62	962	0.44	0.65	0.74	1.09	1.18
17	15,000	Saudi Arabia	N/A	N/A	N/A	13.56	904	0.41	0.62	0.74	1.03	1.15
18	20,000	Saudi Arabia	N/A	N/A	N/A	16.64	832	0.38	0.57	0.66	0.94	1.04
19	30,000	Saudi Arabia	N/A	N/A	N/A	23.64	788	0.36	0.54	0.57	0.89	0.93
20	40,000	Cyprus	N/A	N/A	N/A	28.52	713	0.32	0.49	0.57	0.81	0.89
21	50,000	Cyprus	N/A	N/A	N/A	33.35	667	0.30	0.45	0.56	0.76	0.86

[a]Divided by Q_w not by $90\%.Q_w$;

[b]Equals RO investment cost multiplied by annuity factor from Eq. 3.1 (10 years lifetime, plus infrastructure investment cost multiplied by annuity factor (20 years lifetime), all divided by annual production (90% of Q_w)

[c]Estimated O&M is 60% of unit production cost (GWI, 2006)

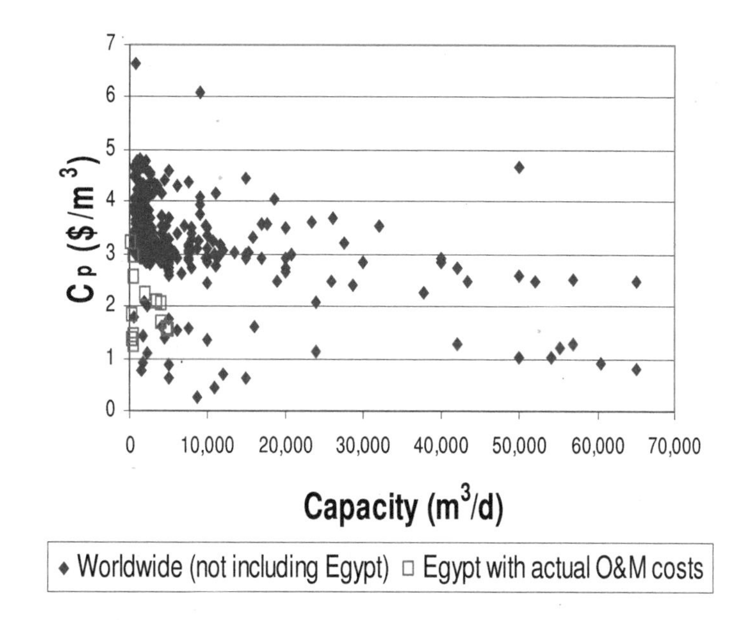

Figure 3.3 RO unit production costs versus plant capacity (worldwide data set from (GWI, 2006) and data from Table 3.1)

Table 3.2 shows a breakdown of O&M unit costs in Egypt and internationally, taking an example from Greece and some developing countries in Asia.

Table 3.2 Comparison between RO desalination O&M unit costs in Egypt and other countries (normalized to 2001 US$)

	Greece (Voivontas *et al.*, 2003)	**Egypt[a] (Abou Rayan *et al.*, 2003)**	**Arid developing countries in Asia[b] (UNEP, 2000)**
Recovery rate (%)	38	30	N/A
Range of plant capacities (m^3/d)	500-10,000	500-5,000	1,200-18,000
Specific energy consumption (kWh/m^3)	5.00	6.5-9.0	8.61
Electricity cost (US$/kWh)	0.11	0.06	0.05
Electricity total cost (US$/$m^3$)	0.55	0.39-0.54	0.45
Chemicals consumption (kg/m^3)	0.28	0.43	N/A
Chemicals cost (US$/kg)	0.18	0.52	N/A
Chemicals total cost (US$/$m^3$)	0.05	0.23	0.04
Membrane replacement rate (%)	12	10	20
Membrane replacement cost (US$/$m^3$)	0.06	0.21	0.19
Labour (US$/$m^3$)	0.18	0.16-0.17	0.04-0.08
Maintenance & repair (US$/$m^3$)	N/A	0.08	0.08-0.16
Other (% of operation cost)	10	N/A	N/A
Total (US$/$m^3$)	0.94	1.08-1.12	0.81-0.93
Cost of electricity (% of total)	59	36-49	48-55

[a]The higher the capacity, the lower the unit labour costs;
[b]The higher the capacity, the lower the unit labour and maintenance costs are.

From Table 3.2, the following observations are made when comparing values of Greece and Egypt:

- Energy consumption for RO plants in Egypt is high but the cost of electricity is much lower;
- More chemicals are used in the case of Egypt;
- The cost of electricity for all cases ranges from 36 to 59% of total O&M costs. Hence, it may be attractive to use solar energy to ensure sustainability and reduce costs (Chapter 4);
- Membrane replacement cost is higher in Egyptian RO plants due to the costs of importing chemicals and membranes.

When comparing RO plants in Egypt to other examples from developing countries in Asia (as shown in Table 3.2), one can conclude that the values are similar for the energy, membrane replacement, maintenance and repair cost components but are higher with respect to the chemicals and labour components.

It can be concluded from Tables 3.1 and 3.2 that O&M costs in Egypt did not follow the estimation of the IDA desalination report (2006), via the 60% of unit production cost value, most likely due to small plant capacities and relatively high chemical and labour costs.

We fitted two equations to calculate unit capital cost and unit production cost (using actual O&M costs) for RO plants in dependence of capacity for plants listed in this paper (21 plants in Egypt and world-wide) as shown in Figures 3.4 and 3.5:

$$C_c = 1340.9\exp^{-2\times10^{-5}Q_w} \quad \text{Equation 3.2}$$

$$C_p = 6.25Q_w^{-0.17} \quad \text{Equation 3.3}$$

where C_c is unit capital cost in US\$/m^3/d; C_p is unit production cost in US\$/m^3 (equal to annual cost of capital plus actual annual O&M costs and both divided by annual output) and Q_w is capacity expressed as flow rate of product water in m^3/d.

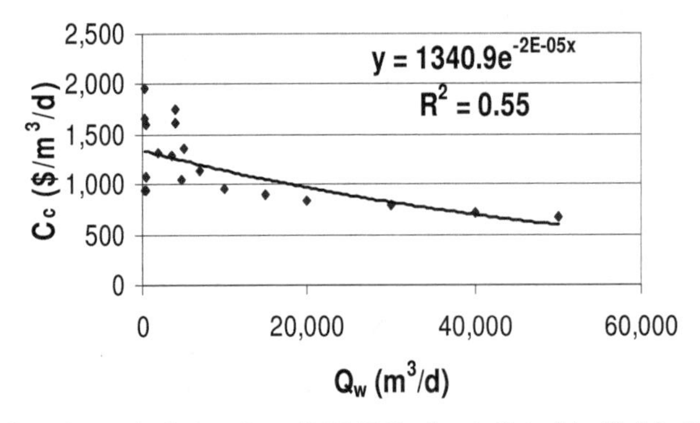

Figure 3.4 RO unit capital cost for all 21 RO plants listed in Table 3.1 (in 2001 US$)

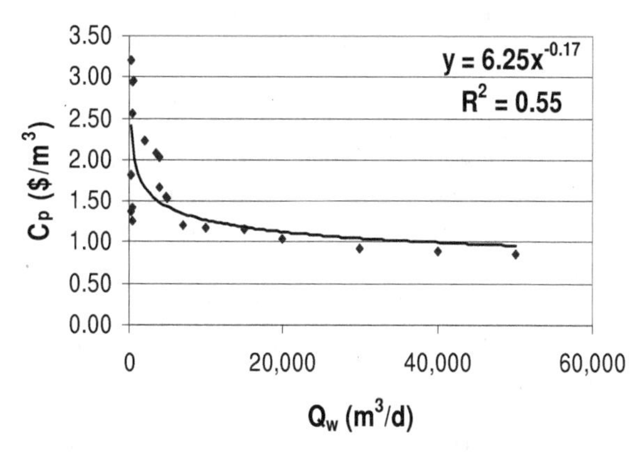

Figure 3.5 Unit production costs for all 21 RO plants listed in Table 3.1 (in 2001 US$)

Costs for future RO plants are mainly dependent on local factors such as energy costs, qualified labour, source and quality of intake water. The equations proposed here are meant as an indication for costs of RO plants in the capacity range of 250 to 50,000 m^3/d for Egypt.

Option 2: Long-distance piping. Basic cost equations

The cost of transporting potable water by pipeline is mainly dependent on distance, elevation difference, soil conditions, labour costs, electricity costs, and spare parts costs. It is therefore difficult to compare pipeline construction costs from one location to another. The most accurate cost estimate is obtained after first carrying out a concept design for the project. In this paper, we nevertheless propose a relationship for unit capital and unit O&M costs as a function of distance and capacity, based on the specific case of pumping treated water from various locations along the Nile branches to the South Sinai region (a maximum distance of 450 km). Nile water is treated through sedimentation, sand filtration and then chlorination before being pumped through long-distance piping to its destination. The soil conditions in this region are mainly sandy soil; there are hills in the area up to 200 m which can be mostly by-passed.

According to our knowledge, there are no publications on the current cost of water transportation by pipelines in Egypt or comparable locations. Zhou and Tol (2005) state that little has been published on costs of transporting water over long distances. Nevertheless, many authors have emphasized that unit production costs of desalination are competitive with those of long-distance water piping (Abou Rayan *et al.*, 2001; WC, 2004). Zhou and Tol (2005) presented a comparison of transportation costs gathered from literature. They mention that the most detailed analysis is that of Kally (1993). However, this publication is based mostly on much older reports. Using Kally (1993), Zhou and Tol (2005) estimated costs of transporting desalinated water inland. In this paper we provide more recent and detailed cost data for long-distance piping.

Kally (1993) divided transportation cost (which is equivalent to unit production cost) into horizontal and vertical distances, with costs of 6.1 cents/m^3 per 100 km horizontal distance, and 5.2 cents/m^3 for lifting the water 100 m (1993 US$, not adjusted to 2001). These costs are for transfer in an open canal in soft but stable soil. Transporting water by pipeline would lead to an increase in cost by 271%, and if the soil is sandy, costs would be higher by further 175% according to Kally (1993). Pipelines are preferred over canals to reduce water losses especially in highly permeable soils (Elarabawy *et al.*, 2000).

A pipeline from Suez to supply Sharm (along with other main cities on the Gulf of Aqabah) with treated Nile water was constructed in three stages of 168, 100 and 100 km respectively in 1986, 1997 and 1998. Table 3.3 shows details of this project together with other long-distance water piping projects in Egypt. Water treatment costs are included as part of the transportation costs for pipes 5 and 7, as the original document showed the investment costs for the treatment and the piping projects as a lump sum. The annual cost of capital is calculated using 30 years lifetime, and a discount rate of 8%. All costs are normalized to the US$ of 2001.

A multiple linear regression (MLR) was performed on the data presented in Table 3.3 to correlate unit capital and unit production costs with length and capacity of the pipelines. The following two relationships were obtained (see Figure 3.6 and Figure 3.7):

$$C_c = 49 + 18.5L - 0.04Q_w \quad \text{Equation 3.4}$$

$$C_p = 0.04 + 0.01L - 10^{-5}Q_w \quad \text{Equation 3.5}$$

where C_c is unit capital cost in US$/m^3/d; C_p is unit production cost in US$/m^3; L is length in km; and Q_w is capacity in m^3/d.

The cost equations for Option 2 assume that the source water requires either no treatment or only minimal conventional treatment (sedimentation, sand filtration and chlorination in the case of the Nile water) and these costs are included in the empirical cost equations. The proposed cost relationship should also hold for other locations with similar characteristics and similar labour costs as the one described above.

If the equations of Kally (1993) are followed (after being normalized to the US$ 2001), then the unit production cost of transporting water by a pipeline in a sandy soil is 1.28 US$/m^3 (distance of 368 km and elevation height taken as 100 m to account for differences in elevation along the way). This figure is lower than the 1.9 US$/m^3 obtained using Equation 3.5 with length (L) equal to 368 km and capacity (Q_w) equal to 15,000 m^3/d.

Comparison of costs

Figures 3.6 and 3.7 show unit capital and unit production costs respectively for both Option 1 (RO desalination) and Option 2 (long distance piping), using Equations 3.2 and 3.3 for Option 1 and Equations 3.4 and 3.5 for Option 2.

These figures show that unit capital costs for those pipelines which are longer than 140 km are above the cost of a possible RO desalination plant at any flow capacity. For unit production cost on the other hand, desalination costs are lower than long-distance piping starting from a pipeline length of at least 300 km and plant capacity of at least 2000 m^3/d. Another example is that for a 350 km distance, desalination costs are lower than long distance piping for a plant capacity of 500 m^3/d or greater.

Table 3.3 Costs of long-distance piping projects in South Sinai, Egypt (normalized to 2001 US$). Data from (Khaled, 2008) except for pipe no. 9 which is taken from (Abou Rayan *et al.*, 2003b)

No.	Length (km) (L)	Start and end points (City)	Type of source water	Diameter (mm)	Capacity (m^3/d) (Q_w)	Year (end of construction)	Investment Cost (million US$)	Unit capital cost (US$/$m^3$/d) ($C_c$)	Annual cost of capital per unit[a] (US$/$m^3$)	Unit O&M costs (US$/$m^3$)	Unit Production cost (US$/$m^3$) ($C_p$)	Project details
	C1	C2	C3	C4	C5	C6	C7	C8=C7/ C5	C9=a.C7/C5	C10	C11=C9+ C10	
1	30	St Catherine to St. Catherine	Potable ground water	N/A	4,000	N/A	3.5	882	0.28	0.09	0.37	Includes 2 pumping stations
2	79	El Tor to El Tor	Potable ground water	N/A	7,000	N/A	6.2	885	0.28	0.03	0.32	Includes storage tank of 1000 m^3 and 1 pumping station
3	100	Abou Rudeis to El Tor	Treated Nile water	430	15,000	1998	19.2	1278	0.31	0.11	0.42	2 pumping stations
4	100	El Tor to Sharm	Treated Nile water	400	15,000	1997	20.5	1363	0.33	0.11	0.44	2 pumping stations
5	168	Suez to Abou Rudeis	Treated Nile water	350/ 605	35,000	1986	28.5	815	0.20	0.18	0.38	Treatment plant, 14 storage tanks, 2 pumping stations
6	168	Suez to Abou Rudeis	Treated Nile water	800	65,000	Under construction	73.5	1130	0.28	0.18	0.46	Parallel extension to line 6, 2 pumping stations with storage tanks
7	360	Suez to Ras Naqb	Treated Nile water	1000	65,000	Under consideration (since 1999)	235	7646	0.88	0.50	1.38	Treatment plant, 12 pumping stations, storage tanks
8	450	El Koraimat to Hurgada	Treated Nile water	600/ 1000	28,000	1997	214	3608	1.86	0.49	2.35	N/A

[a]Equals C7 multiplied with annuity factor a (from Eq. 3.1) and divided by Q_w

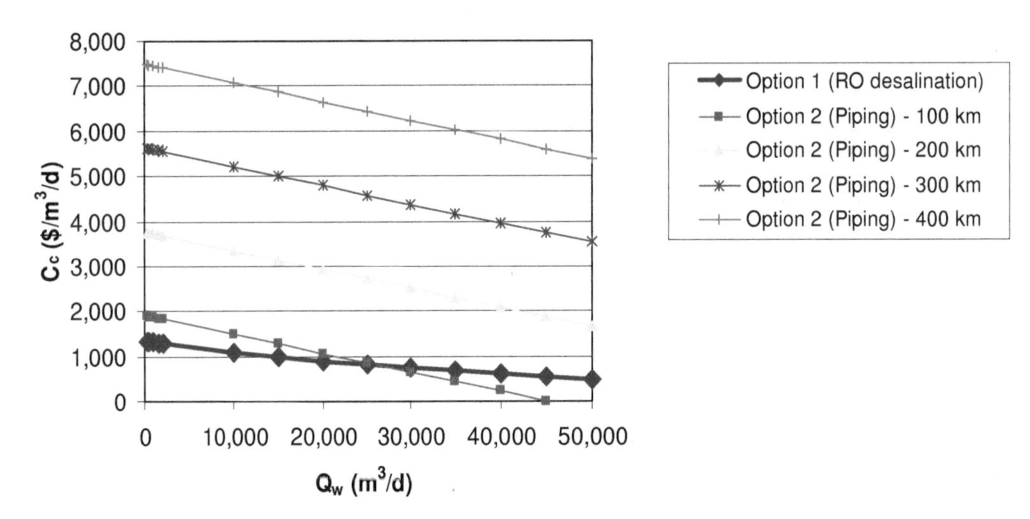

Figure 3.6 Unit capital costs (C_c) of RO desalination and long distance piping versus capacity and pipe length

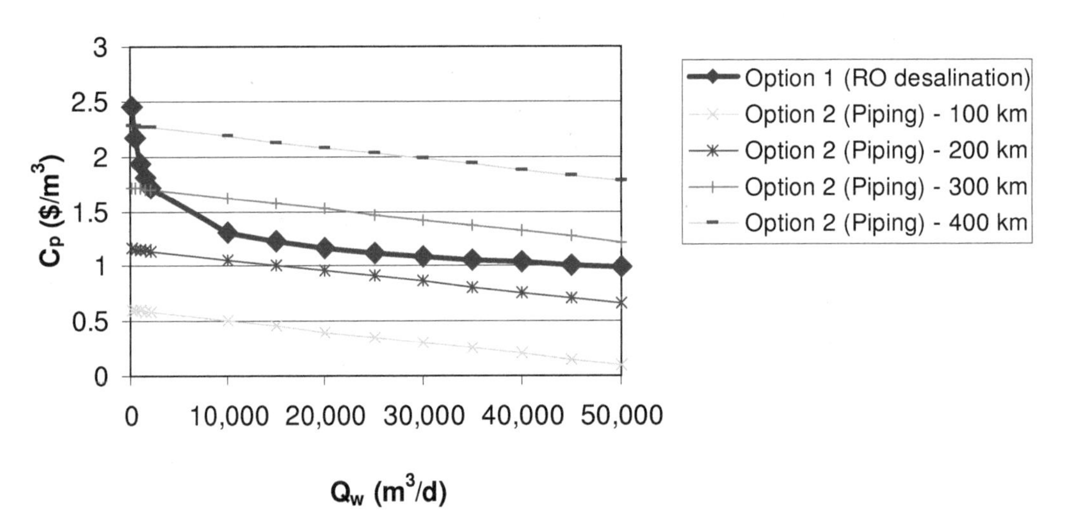

Figure 3.7 Unit production costs (C_p) of RO desalination and long distance piping versus capacity and pipe length

Depending on capacity and length of the pipeline, RO desalination can be comparable and even cheaper to long-distance piping. The real cost of long-distance piping is expected to be higher if the cost of Nile water is factored in. However, it needs to be pointed out that RO desalination can have some negative environmental impact due to high energy consumption per m^3 production and brine containing high salt concentrations and chemicals.

Conclusions

This paper provides an indication of the costs of RO desalination and long-distance piping in Egypt and elsewhere. Both options are meant to provide water of potable quality. The cost equations for Option 2 assume that the source water requires either no treatment or only minimal conventional treatment (sedimentation, sand filtration and chlorination in the case of the Nile water) and these costs are included in the empirical cost equations.

Unit production costs of RO plants in Egypt are comparable to plants of similar capacity (above 600 m^3/d) in the IDA desalination report (2006). Earlier observation that RO desalination costs in Egypt are higher than international trends (Hafez and El Manharawy, 2002) is not valid as plant capacities were not taken into consideration (Egyptian RO plants being generally smaller than 5,000 m^3/d). However, it should be noted that current standard industry unit production costs are much lower (about 0.5 to 1 US$/$m^3$) than quoted figures in this paper. Most of the quoted desalination plants in this chapter are quite old (they were built in the 1990's), they are small in size, and recently specific energy consumption has been remarkably reduced by the use of isobaric pressure exchange - energy recovery system which could also be applicable for small RO systems.

Actual O&M costs are higher in Egypt than the estimated O&M costs (taken as 60% of unit production cost as suggested in (GWI, 2006)). It is important to consider the O&M costs relative to capacity/output of the plant when estimating costs of future desalination plants.

Basic empirical cost equations were proposed to estimate RO costs and costs of long-distance piping in Egypt. The boundary of applicability for the RO cost equations is a plant capacity between 250 m^3/d to 50,000 m^3/d; and for the long distance piping equations a pipe capacity up to 65,000 m^3/d, length up to 450 km, and soil and ground conditions similar to South Sinai.

RO desalination can be cheaper than long distance piping depending on capacity (of RO plant or pipeline) and length of pipeline. However, RO desalination can have some negative environmental impacts which have to be examined before considering the construction of new plants. More sustainable water supply options such as reuse of treated wastewater and water demand management should also be considered for the development of South Sinai.

Acknowledgements
We would like to thank Prof. Magdy Abou Rayan, Mansoura University, Egypt and Mr. Ibrahim Khaled, Sinai Development Authority, Egypt for the critical review of this paper and for kindly providing valuable data, respectively.

4 Impact of solar energy cost on water production cost of seawater desalination plants in Egypt

Abstract

Desalination is performed using electricity, mostly generated from fossil fuels with associated greenhouse gas emissions. Increased fuel prices and concern over climate change are causing a shift to alternative sources of energy such as solar energy.

An equation is proposed to estimate the unit production costs of reverse osmosis (RO) desalination plants as a function of plant capacity, price of energy and specific energy consumption. This equation is used to calculate unit production costs for desalinated water using photovoltaic (PV) solar energy. Unit production costs of desalination plants using solar energy are compared to conventionally-generated electricity considering different prices for electricity.

Solar energy is not cost competitive at the moment but advances in the technology will continue to drive the prices down, whilst penalties on usage of fossil fuel will increase electricity costs from conventional non-renewable sources. Solar thermal is cheaper (at a current price of 0.06 US$/kWh) than photovoltaic (PV), however, PV is more appropriate for Egypt as it is more applicable to the smaller RO plant sizes found in Egypt. We would expect that there will be a shift towards more centralized RO plants in Egypt and this would then favour the adoption of solar thermal energy in the near future.

Keywords: solar energy; desalination; unit production costs.

This chapter is based on:
Lamei, A., van der Zaag, P. and von Münch, E. (2008b) Impact of solar energy cost on water production cost of seawater desalination plants in Egypt. *Energy policy*, **36**, 1748-1756.

Introduction

Many arid countries are now considering desalination of seawater or brackish water as an important source of water. Membrane processes, mainly reverse osmosis (RO), are currently the fastest growing technologies in water desalination (Semiat, 2000), but they are energy intensive. More than 30% of unit production cost of RO desalinated water is attributed to energy costs (Hafez and El Manharawy, 2002). Therefore, a small change in energy prices (measured in US$/kWh) can affect the unit production costs of RO plants.

Although many arid countries especially in the Middle East have an abundance of year-round solar radiation, most RO desalination plants worldwide currently use conventional non-renewable energy sources with associated carbon dioxide emissions contributing to global climate change. For each 1 m^3 of water generated in an RO plant, using fossil fuel (oil), about 3 kg of CO_2 are produced (Fiorenza *et al.*, 2003).

As Egypt has signed the Kyoto protocol in 1999, efforts are being made to reduce the per capita generation of Carbon. The Kyoto protocol requires the global per capita emissions to drop to 0.2-0.7 ton C/cap/yr compared to the current value of 0.3 in developing countries (0.43 for Egypt) and to 5.5 from USA and 2.5 from Western Europe. This requires industrialized countries to drop by a factor of 10 or more the per capita emissions and at the same time, leaves no margin for increased emissions in many developing countries. Rather, a radical transformation of the energy system is required (Sanden and Azar, 2005; Unknown, 1998). In 2005, the value of carbon dioxide emissions for Egypt was approximately 2 ton CO_2/cap/yr and the total CO_2 emissions from the consumption and flaring of fossil fuels were 135 million tons per year (EEDRB, 2005).

The total capacity of desalination plants in Egypt, mainly RO, is approximately 130,000 m^3/d (GWI, 2006), or about 1% of all domestic water usage (based on 75 million inhabitants and 120 l/cap/d water usage; this figure excludes industrial and agricultural water use). These RO plants produce about 142,000 tons of CO_2 per year. This figure was calculated assuming that the current typical specific energy consumption of RO plants is 5 kWh/m^3 although it can be even up to 9 kWh/m^3 due to increased salinity of the intake seawater and lower energy efficiency of plants. The total amount of produced CO_2 will increase further as more water is sourced from desalination in Egypt in the future. If Egypt was to produce all of its domestic (non-industrial, non-agricultural) water demand from RO desalination using fossil fuel (oil), this would result in about 10 million tons of CO_2 emissions per year or 7.4% of total CO_2 emissions from the consumption and flaring of fossil fuels – which shows that this scenario would be entirely undesirable from a climate protection point of view.

To reduce the CO_2 production from RO plants, renewable energy sources, such as solar-generated power, could be used. Solar energy is available when demand for water is high (during the day time during summer), which is advantageous. Egypt has more than 3,600 hours of sunshine per year, compared to 3,000 hours in the Mediterranean region and only 1,600-2,400 hours in Northern Europe (Hoffmann, 2006). This type of decentralised energy generation could also be important for remote areas where it is not economical to extend an electrical grid.

The main purpose of this paper is to investigate whether solar energy can be a cost effective energy source for desalination plants for water production in arid countries like Egypt. The paper discusses costing aspects of two systems for solar energy generation: i) photovoltaic (PV) systems and ii) solar thermal systems.

An empirical cost equation is proposed to calculate unit production costs for RO desalination plants as a function of plant capacity, energy cost and specific energy consumption. All prices are quoted in US$ (2006).

Technologies for desalination and solar energy generation

Desalination Technologies

Table 4.1 shows available desalination technologies. Thermal technologies have higher total energy consumption than membrane technologies for desalination (13 kWh/m^3 compared to 5 kWh/m^3 respectively) (GWI, 2006). However, thermal desalination plants can make use of the waste heat of power stations which can reduce their energy demand. Another advantage of thermal desalination is that its performance is unaffected by feed water salinity. For high salinity feed water, membrane technologies have a lower recovery rate and hence higher specific energy consumption. Another aspect in favor of thermal desalination is that large scale RO plants can cause pollution due to chemicals added to the brine rejects (hydrochloric or sulfuric acid, anti-scalants, etc.) which are disposed into the sea, whereas this is not the case for thermal desalination processes.

Table 4.1 Available desalination technologies

Thermal desalination	Membrane
Multistage flash (MSF)	Reverse osmosis (RO)
Multiple Effect distillation (MED)	Electrodialysis reversal (EDR)
Vapour compression (VC)	

Therefore, when seawater desalination on a large scale is required, thermal desalination plants will continue to be preferred to produce water from a water source with a high salinity concentration. In order to enhance energy efficiency, such plants are usually coupled with thermoelectric power stations according to a dual purpose scheme (generation of both water and electricity).

However, the implementation of large centralised thermal desalination plants is still faced with logistic problems, whereas the availability of robust RO modules facilitates its implementation (Fiorenza *et al.*, 2003). Also, the coupling of thermal desalination with power stations is not always feasible in terms of location and availability. For this reason, the market for RO will continue to grow in the future. RO plants will occupy the market for small to medium plants or even big plants in a single purpose system (water production only).

Desalination with Solar Energy

Solar energy is produced by either photovoltaic (PV) or solar thermal systems. Photovoltaic systems convert solar energy directly into electricity. A PV cell consists of two or more thin layers of semi-conducting material (silicon). When the silicon is exposed to light, electrical charges are generated and this is conducted away as direct current. The electrical output from a single cell is small, so multiple cells are connected together to form a module. Any number of modules can be connected together to give the desired electrical output (Figure 4.1, left).

For solar thermal systems, parabolic or flat mirror collectors are used to concentrate solar energy in order to heat feed water so it can be used in the high temperature end of a standard thermal distillation process (Figure 4.1, right).

Figure 4.1 PV modules (left) and parabolic solar collectors for solar thermal systems (right)

Table 4.2 summarizes the main differences between the solar thermal and PV solar energy production technologies. Due to the abundance of solar energy in the Mediterranean and the Middle East regions, many plants have been built which utilize solar energy for water desalination (see Table 4.3 and Table 4.4). The most common combination of solar energy and desalination in Egypt is PV-RO. However, its usage is still limited to small/pilot scale applications in remote areas (except for the Red Sea plant which is an old one). Most of the other Arab Gulf countries depend on thermal desalination which can be coupled directly with solar thermal energy (Table 4.4).

Table 4.2 Comparison between solar thermal and photovoltaic systems (PV)

	Photovoltaic (PV)	Solar thermal
Principle	Converts solar energy into electric charges	Concentrates solar energy into heat energy
Use with RO desalination	Direct	Need intermediate phase (steam turbines)
Use with thermal desalination	Not applicable	Direct
Scale of application and capacity	Small to medium scale Max. 5 MW (current) (corresponding to desalination plant capacity of 10,000 to 15,000 m^3/d)	Large scale only Starts from 5 MW, 345 (current) to 600 MW (future)

Water and energy needs in Egypt

Water Production from RO Plants in Egypt

Egypt (with a current population of 78 million) (CIA, 2005) is among those countries which are very vulnerable with regard to water scarcity. Figure 4.2 shows the per capita availability of water and increasing population up to the year 2025. One of the options to address water shortages is RO desalination of seawater, which is now the prevailing desalination technology in Egypt (Abou Rayan *et al.*, 2001). The unit production cost of RO desalinated water in Egypt ranges from 1 to 3.5 US$/m^3 depending on plant capacity. The cost of energy constitutes up to 33% of the unit production costs for RO plants (Chapter 3). The specific energy consumption of RO plants in the region is typically 6.5-9 kWh/m^3 depending on the salinity of the intake water (seawater) and the age, efficiency and configuration of the plant.

Table 4.3 Desalination plants in the Mediterranean and Middle East regions using PV systems as a source of energy (Abu Arabi, 2004; Al Suleimani and Nair V., 2000; Karagiannis *et al.*, 2007)

Plant location, commissioning year	Intake water type (salinity in ppm)	Desalination process	Plant capacity for product water (m^3/d)	Water production cost (US$/m^3)	Comments
Jeddah (Saudi Arabia), 1981	SW [a] (42,800)	RO	3.25	Unknown	
Doha (Qatar), 1984	SW	RO	5.7	Unknown	
Red Sea (Egypt), 1986	BW [b] (4,400)	RO	240	Unknown	
Algeria, 1987	BW (3,200)	RO	22.8	Unknown	
Lampedusa (Italy), 1990	SW	RO	120	9.75 (in year 2004)	
Oman, 1998	BW	RO	5	6.25(in year 2004)	
Turkey, 2007	BW	RO	3	Unknown	Tendering phase
Morroco, 2007	BW	RO	8	Unknown	Tendering phase

[a] SW=seawater; [b] BW=brackish water

Table 4.4 Desalination plants in the Mediterranean and Middle East regions using solar thermal systems as a source of energy (Abu Arabi, 2004; Karagiannis *et al.*, 2007)

Plant location, commissioning year	Intake water Type	Desalination process [b]	Plant capacity (m^3/d)	Water production cost (US$/m^3)
Kuwait, 1980	SW[a]	MSF	100	Unknown
Kuwait, 1983	SW	MSF	10	Unknown
Abu Dhabi (UAE), 1985	SW	MED	120	8 (in year 2004)
Kuwait, 1987	SW	MSF RO	250 20	Unknown
Arabian Gulf (unknown), 1987	SW	MED	6000	Unknown
Dead Sea (Jordan) 1987	SW	MED	3000	Unknown
Almeria (Spain), 1988	SW	MED	72	5 (in year 2004)
Al Ain (UAE), 1991	SW	MSF	500	Unknown

[a] SW=seawater , [b] abbreviations explained in Table 4.1

Conventional Energy Generation in Egypt

In Egypt, energy for desalination is obtained either from the national electricity grid (generated using oil or natural gas) with a price of 0.06 US$/kWh, or from diesel generators (in remote locations) at a price of 0.07 US$/kWh. The world-wide price of electricity ranged from 0.05 to 0.09 US$/kWh in 2003 with the lower range for the Gulf countries and the upper range for Europe and the United States (Abou Rayan *et al.*, 2003).

Egypt has an installed electricity generation capacity of 17 Gigawatt (GW) and plans to considerably increase this capacity by 2010. Around 84% of Egypt's electricity capacity consists of thermal turbines powered by natural gas. The proportion of natural gas in the electricity sector is expected to increase after the recent discovery of further reserves of natural gas. Under a recently completed program, all oil-fired power plants have been converted to operate on natural gas in Egypt. The demand for electricity has rapidly grown in recent years (7-8% per year) due to the industrial development of the country. Egypt intends to build additional thermal power plants (Youssef, 2006).

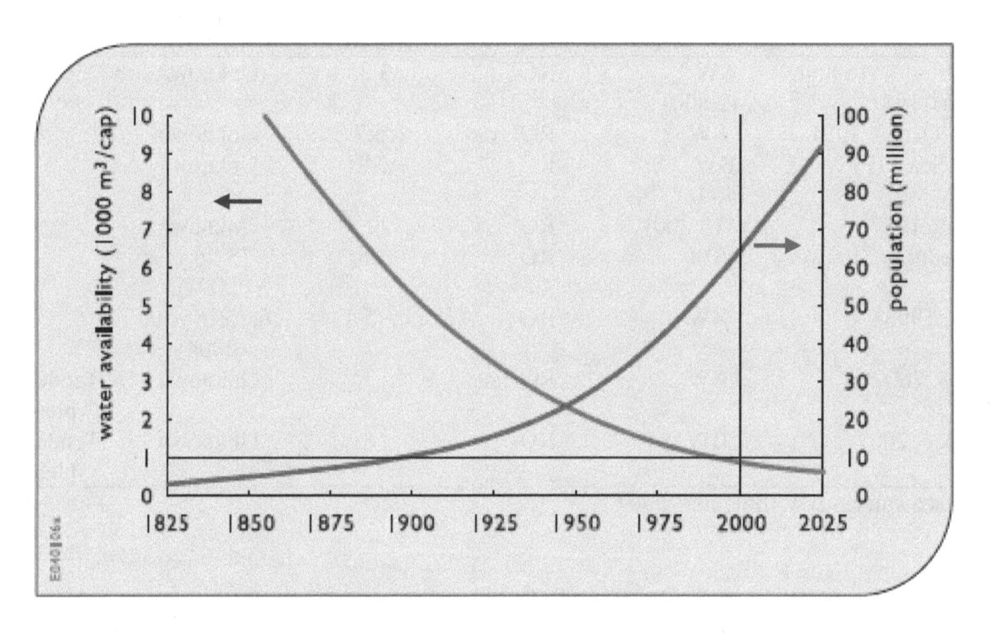

Figure 4.2 Water availability and population up to 2025 in Egypt (MWRI, 2005)

Solar Energy Use in Egypt

Solar energy is used in Egypt for the electricity supply of isolated areas (using PV technology). In addition, the government is also planning to build a power plant which will generate 30 MW of solar energy capacity out of 150 MW of total capacity. The project will be funded by the Global Environment Facility (GEF), which will cover the cost difference between solar and thermal capacities (Youssef, 2006) .

The perspectives for the development of the co-generation sector (generating electricity using both solar and natural gas) are also good, given the increased usage of natural gas in the industrial sector. Co-generation could significantly reduce the energy costs for energy intensive industries. But the main problem for the co-generation development in Egypt is the low price of natural gas. The price of natural gas is subsidized by the government and there is therefore little incentive for companies to invest in costly co-generation power plants (CEEI, 2001). As far as we know, the only desalination plant in

Egypt which uses PV technology for energy generation is the old RO plant desalinating brackish water near the Red Sea (Table 4.3).

As far as we are aware, the only other use of solar energy to produce water in Egypt is performed using solar stills. Solar stills use solar energy directly, unlike PV or solar thermal which require a medium to convert solar energy to either electric or heat energy, respectively. Solar stills are shallow basins, lined with black plastic liners which hold seawater and absorb solar radiation. Water vapor condenses on the underside of a sloped transparent cover, which then runs through troughs and is collected in tanks at the end of the still. Average water production from solar stills is about 4 $l/m^2/d$ (Chaibi, 2000). Solar stills are used to supply fresh water to isolated communities which are far from any other fresh water source.

Although Egypt has a solar intensity of about 3000 $kWh/m^2/yr$ (El Nokrashy, 2005), the availability of natural gas at low prices makes solar energy not competitive at the current prices.

RO desalination using conventional versus solar energy

RO Desalination Unit Production Cost

An empirical basic cost equation was previously developed to calculate the unit production cost ($US\$/m^3$) for RO desalination plants as a function of plant capacity. Data were obtained from 21 RO plants located in the Mediterranean and Red Sea regions (Chapter 3):

$$C_p = 6.25 Q_w^{-0.17} \qquad \text{Equation 3.3}$$

where C_p is unit production cost in $US\$/m^3$ and Q_w is plant capacity in m^3/d. This equation is valid for RO plants in the capacity range of 250 to 50,000 m^3/d for the Mediterranean and Red Sea regions (Table 4.5).

In this paper, a multiple-linear regression (MLR) analysis (Figure 4.3) was performed to derive a cost equation as a function of plant capacity and cost of energy (overall correlation was 73%):

$$C_p = 1.6 - 3.42 \times 10^{-5} Q_w - 9.76 p + 0.14 f_e \qquad \text{Equation 4.1}$$

where C_p is unit production cost in $US\$/m^3$; Q_w is plant capacity in m^3/d; p is price of energy in US\$/kWh; f_e is a factor for specific energy consumption in kWh/m^3. Equation 4.1 is only valid for values of Q_w in the range of 250 to 50,000 m^3/d, p in the range of 0.05 to 0.07 US\$/kWh and f_e in the range of 5 to 15 kWh/m^3 (see Table 4.5).

Figure 4.4 shows unit production costs of 21 RO desalination plants located in the Mediterranean and Red Sea regions. Figure 4.4 also shows the calculated effect of energy prices on unit production costs of RO desalination. Doubling the electricity prices would increase the RO unit production costs within a range of 12 to 40% irrespective of plant capacity. For the surveyed 21 RO plants, the sensitivity of the unit production costs to an increase in electricity prices depends on their specific electricity consumption f_e which is not influenced by RO plant capacity but by the energy efficiency of the plant, age of the plant (the older the plant, the higher the specific energy consumption) and whether there is an energy recovery system or not. Table 4.5 provides details for each of these RO plants.

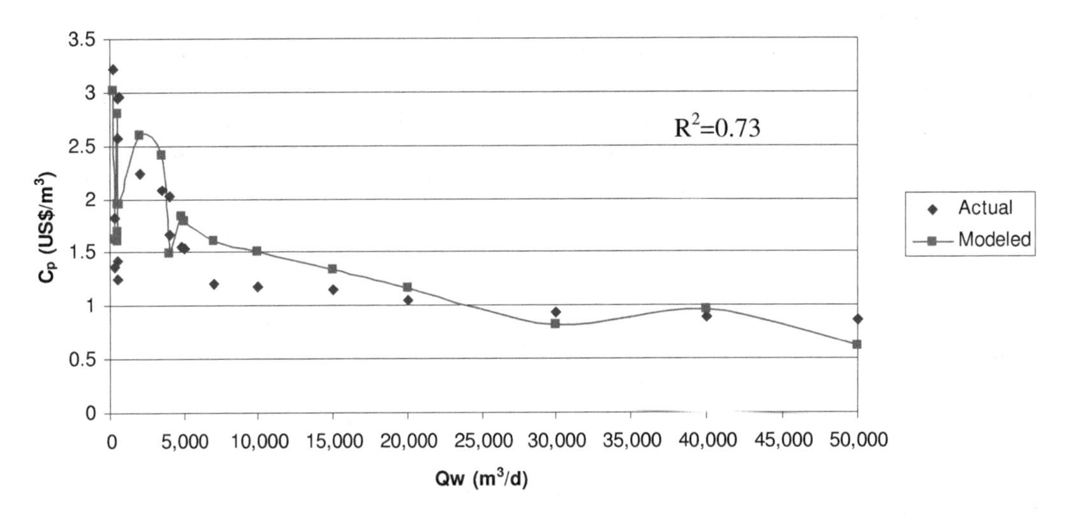

Figure 4.3 Multiple linear regression (MLR) for unit production costs C_p of RO desalination versus plant capacity Q_w (Mediterranean and Red Sea regions) with different energy costs and energy consumption values (Table 4.5)

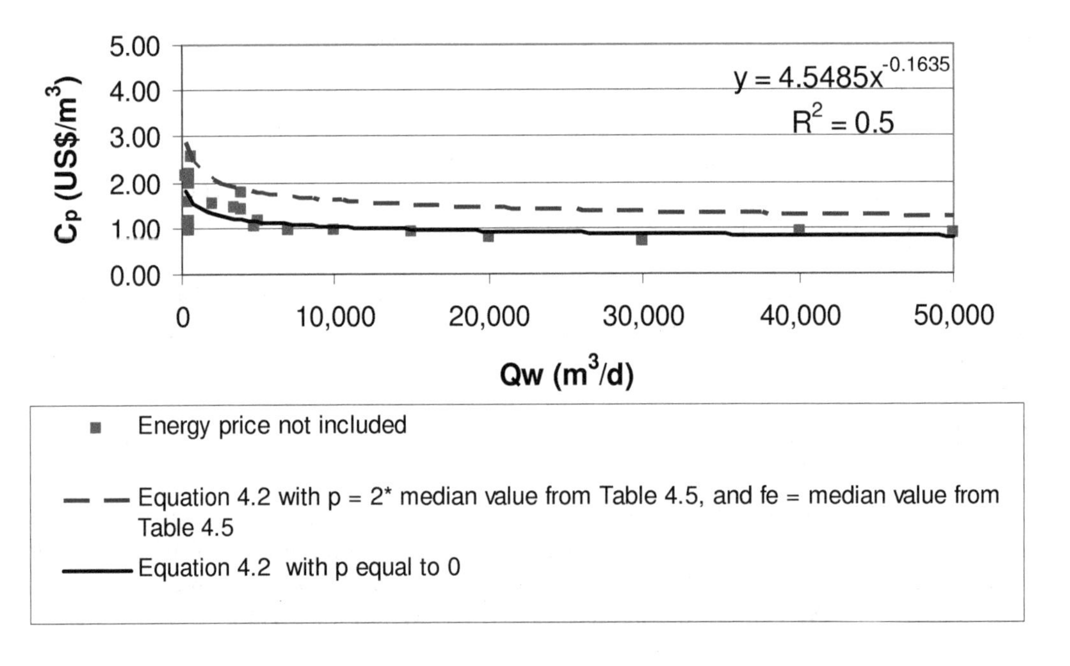

Figure 4.4 Unit production costs of RO desalination versus plant capacity (Mediterranean and Red Sea regions) with different energy costs

Table 4.5 Details of 21 RO plants located in the Mediterranean and the Red Sea regions

No.	Location	Capacity Q_w (m^3/d)	Unit production cost C_p (US$/$m^3$)	Energy-related unit production cost (US$/$m^3$)	Energy price p (US$/kWh)	Specific energy consumption f_e (kWh/m^3)	References
1	Hurgada (Egypt)	250	3.21	1.04	0.07[a]	15	(Hafez and El Manharawy, 2002)
2	Nuweiba (Egypt)	300	1.82	0.25	0.06	4	(Khaled, 2008)
3	Red Sea (Egypt)	350	1.36	0.25	0.06	4	(Abou Zeid, 2006)
4	Safaga (Egypt)	500	2.94	0.94	0.07[a]	13	(Hafez and El Manharawy, 2002)
5	Hurgada (Egypt)	500	1.42	0.25	0.06	4	(Abou Rayan *et al.*, 2003)
6	Matrouh (Egypt)	500	1.25	0.28	0.06	5	(Abou Rayan *et al.*, 2003)
7	Dahab (Egypt)	500	2.57	0.40	0.06	7	(Abou Rayan *et al.*, 2001)
8	Taba (Egypt)	600	2.95	0.40	0.06	7	(Abou Rayan *et al.*, 2001)
9	El-tor (Egypt)	2000	2.23	0.70	0.06	12	(Hafez and El Manharawy, 2002)
10	Sharm (Egypt)	3,500	2.08	0.64	0.06	11	(Hafez and El Manharawy, 2002)
11	Gulf of Aqaba (Egypt)	4,000	1.67	0.25	0.06	4	(Abou Rayan *et al.*, 2003)
12	Sharm (Egypt)	4,000	2.02	0.25	0.06	4	(Abou Rayan *et al.*, 2001)
13	Hurghada (Egypt)	4,800	1.55	0.52	0.07[a]	7	(Hafez and El Manharawy, 2002)
14	Hurghada (Egypt)	5,000	1.54	0.40	0.06	7	(Abou Rayan *et al.*, 2003)
15	Libya	7,000	1.21	0.25	0.05[c]	5	(Hafez and El Manharawy, 2002)
16	Tunis	10,000	1.18	0.25	0.05[c]	5	(Hafez and El Manharawy, 2002)
17	Saudi Arabia	15,000	1.15	0.25	0.05[c]	5	(Hafez and El Manharawy, 2002)
18	Saudi Arabia	20,000	1.04	0.25	0.05[c]	5	(Hafez and El Manharawy, 2002)
19	Saudi Arabia	30,000	0.93	0.25	0.05[c]	5	(Hafez and El Manharawy, 2002)
20	Cyprus[b]	40,000	0.89	-	0	5	(Hafez and El Manharawy, 2002)
21	Cyprus[b]	50,000	0.86	-	0	5	(Hafez and El Manharawy, 2002)
	Average	N/A			0.05	6.5	
	Median (50%-ile)	N/A			0.06	5	

[a] Source of energy for these plants is diesel generators

[b] For these two plants, energy was supplied by the government for free

[c] Price of energy was assumed according to the trend in these countries as it was not explicitly mentioned in the original reference

With increasing size of RO desalination plant, the unit production costs decrease. This favours the adoption of large-scale desalination plants for any type of energy source.

The MLR Equation 4.1 is useful for modeling the measured unit production cost values of 21 plants and showing the relative importance of the three different model parameters (plant capacity, energy price and specific energy consumption). However, this equation cannot be used for extrapolations beyond its range of validity. For this reason, we developed a second equation, which is a power fit for the data of 21 plants, but this time using constant values for p and f_e (the median values as shown in Table 4.5). The resulting equation is:

$$C_p = 4.55 Q_w^{-0.1635} + p \times f_e \qquad \text{Equation 4.2}$$

where C_p is unit production cost in US\$/m^3; Q_w is plant capacity in m^3/d; p is price of electricity in US\$/kWh (median value of 0.06 used); f_e is a factor for specific electricity consumption in kWh/m^3 (median value of 5 used).

Solar Energy Costs: PV and Solar Thermal

The price of both types of solar energy systems (PV and solar thermal) has dropped considerably from 1990 until 2006 (Hoffmann, 2006) and is expected to drop further over the next few years. An opposite trend can be expected for the cost of electricity generated from fossil fuels (0.05 – 0.09 US$/kWh) (Abou Rayan *et al.*, 2003). This is due to the foreseen price rise of fossil fuel as well as the plausible introduction of fiscal penalties for power plants emitting carbon dioxide (Sanden and Azar, 2005).

The solar energy industry expresses the price for solar energy modules as "price per Watt peak (W_p)", which is the number of Watts output under standard conditions of sunlight. For the case of Egypt, one kW_p produces approximately 2000 kWh/yr (equivalent to working about 25% of the time or 6 hr/d) based on the prevailing conditions in the Middle East (US, 2006).

To convert the price per Watt peak (US\$/W_p) into a unit electricity price (in US$/kWh) we used the following procedure: First we calculated the required electricity demand of a given RO plant (e.g. an RO plant with capacity of 10,000 m^3/d and specific electricity demand of 3 kWh/m^3, requires 30,000 kWh/d; then we multiplied that figure with 25% to account for the fact that PV operates for 25% of the time with no batteries used for storing solar energy; the resulting figure is 7500 kWh/d or 1368 kW_p – using the above-mentioned factor of 2000 kWh/year from one module in Egypt). If the price of one module (one W_p) is 1 US\$, then the capital cost for the solar power installation for this plant is 1.4 Million US\$. The annual cost of capital can then be calculated by multiplying the investment cost with the annuity factor, a (Equation 3.1).

$$a = \frac{i \times (1+i)^n}{(1+i)^n - 1} \qquad \text{Equation 3.1}$$

where i is the discount rate (we used 8 %) and n is the economic plant life (we used 25 years lifetime for the PV system). The unit production cost for electricity (in US$/kWh) is then calculated as the annual cost of capital (US$/yr) plus annual O&M costs (US$/yr), divided by the annual plant output (kWh/yr).

The partial operation of the PV system (25% of the day) eliminates the need for a storage battery, therefore reducing O&M costs; O&M costs are considered as 2% of investment cost (Dakkah *et al.*, 2003). Using the above example, it means that 1 US$/$W_p$ is equivalent to 0.06 US$/kWh.

For PV systems, the price (or unit production cost) for the high power band (>70 W_p) solar modules has dropped from 27 US$/$W_p$ (1.53 US$/kWh) in 1982 to a range between 4-8 US$/$W_p$ (0.23-0.45 US$/kWh) in 2006 (including installation) (Dunlop *et al.*, 2001). The industry is aiming to drive down the prices over the next few years to 2 US$/$W_p$ (0.1 US$/kWh) in the year 2010, and 1 US$/$W_p$ (0.06 US$/kWh) in the year 2020 (Hoffmann, 2006).

The price of 1 US$/$W_p$ (or 0.06 US$/kWh) could be achieved by 2020 due to advances in technology on the one hand and/or due to subsidies from the government on the other hand. Also governments can get lower prices for solar modules than individual buyers due to purchasing in bulk. Another factor to reduce price is buying directly from the producer with no distribution channels in between (Dunlop *et al.*, 2001).

The cost of solar thermal energy is 0.06 and 0.09 US$/kWh for parabolic or flat mirrors, respectively (El Nokrashy, 2006), which is already similar to conventionally-generated electricity. However, their application is still limited as they are only suitable for large scale applications (larger than 5 MW; see Table 4.2). Table 4.6 presents a comparison between prices for different types of energy.

Economic Analysis of RO Desalination with PV and conventional energy

As shown in Table 4.6, the cost of solar thermal energy is already comparable to electricity from fossil fuel. The analysis in this paper focuses on comparing electricity generated from PV systems with electricity generated from fossil fuel in order to show the possible economic potential of using PV systems.

Table 4.6 Indicative prices of electricity produced from different types of energy sources

Energy source	1980s	Current (2006)	Future (approx. 2020)
Fossil fuel	< 0.05 US$/kWh	0.05 – 0.09 US$/kWh	> 0.09 US$/kWh (?) (Abou Rayan *et al.*, 2003)
PV	Not known	0.5 US$/kWh	0.06 US$/kWh (Hoffmann, 2006)
Solar thermal	Not known	0.06 – 0.09 US$/kWh	No estimate

For photovoltaic systems, Figure 4.5 shows that solar energy (PV) at current prices (8 US$/$W_p$ which is equivalent to 0.45 US$/kWh) is more expensive than using oil/natural gas for electricity generation. However, at a price of 1 US$/$W_p$, (equivalent to 0.06 US$/kWh), solar energy (PV) can become competitive especially in countries with high conventionally-generated electricity prices (0.09 US$/kWh).

Specific energy consumption rates for desalination plants are decreasing continuously due to advances in technology and use of energy recovery devices. The value for specific energy consumption can be as low as 3 kWh/m^3 for new plants. For that reason, we used that number for Figure 4.5 for the estimation of unit production costs for different energy costs and types.

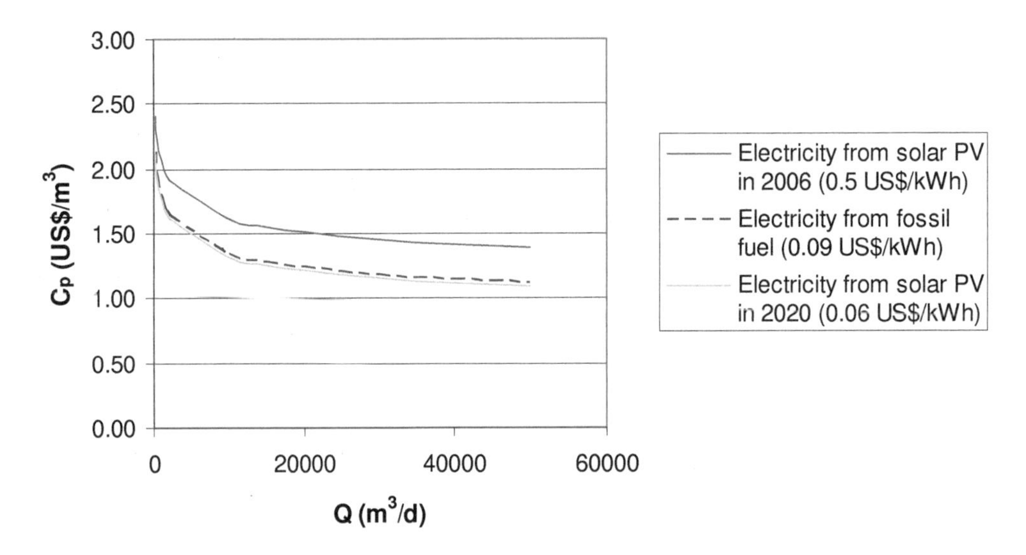

Figure 4.5 Unit production costs of RO plants C_p versus plant capacity Q_w for different energy costs (using Equation 4.2 with f_e equal to 3 kWh/m^3)

Conclusions

Desalination can be performed using thermal or membrane processes (reverse osmosis). RO desalination is the prevailing desalination technology in Egypt using conventionally-generated electricity (from oil or natural gas). The low prices of natural gas in Egypt discourage the use of renewable energy. However, as Egypt has joined the Kyoto agreement, efforts are being made to reduce the per capita CO_2 emissions. In this paper, the focus was on solar thermal and solar photovoltaic (PV) systems to generate electricity for RO plants.

An empirical equation is proposed to calculate unit production cost of RO desalination in dependence of plant capacity, cost of energy and specific electricity demand. Doubling the electricity prices would increase the RO unit production costs within a range of 12 to 40% irrespective of plant capacity. The larger RO plants in our comparison of 21 plants in the Middle East region did not show any economies of scale in terms of annual electricity costs; the plants' energy consumption was instead mainly affected by the energy efficiency of the plant, intake water salinity, age of the plant (the older the plant, the higher the specific energy consumption) and whether there is an energy recovery system or not.

The most common combination of solar energy and desalination in Egypt is PV-RO. However, its usage is still limited to small/pilot scale applications in remote areas. Most of the other Arab Gulf countries depend on thermal desalination which can be coupled directly with solar thermal energy.

When comparing the costs of solar thermal to PV, PV systems are more costly than solar thermal systems at the moment. In fact, solar thermal costs are competitive with local energy prices and can be readily utilized. However, PV can be applied for small to medium projects (RO plant capacity of up to 15,000 m^3/d) and is more popular in Egypt for that reason. It can also be readily connected to an RO desalination plant easing its implementation while solar thermal requires a transitional phase (steam turbines to convert heat energy into electricity).

It appears that due to reducing prices for solar energy generation modules, there is a realistic potential to make RO plants with solar energy use a viable option for Egypt and other countries in North Africa and the Middle East in the future. In Egypt, if the government is to shift towards centralized desalination plants in the future, to tackle increasing water shortage, this would favour the adoption of solar thermal as a more economic and climate-protection friendly energy source.

5 A model for calculation of water demand by the tourism industry in arid coastal regions: The case of Sharm El Sheikh, Egypt

Abstract

In this paper, the water demand by the tourism industry in arid coastal regions, i.e. hotels and related services, is analysed. A model is developed to calculate time-variant water demand by the tourism industry. The model will be incorporated in a comprehensive technical-economic modeling tool for integrated water resources management. The model provides investors (public and private sectors) in water resources projects with the means to estimate current and future water demand required by existing or new tourism facilities. Also the impact of introducing demand management to reduce water consumption in hotel facilities is considered.

Keywords: Model; water demand; arid coastal regions.

This chapter is based on:

Lamei, A., von Münch, E., Imam, E., and van der Zaag, P. (2006) A model for calculation of water demand by the tourism industry. *Integrated Water Resources Management Conference and Challenges of the Sustainable Development (Moroccan Committee of the International Association of Hydrogeologists)*, Marrakech, Morocco, 23-25 May.

Introduction

Water is a limiting resource in coastal zones characterized by high population density, intense economic activity and tourism, with a water demand that is seasonally fluctuating. The usual way to tackle water demand is through conventional surface and ground water abstraction. However, due to increasing limitation in water resources, a shift is taking place towards non-conventional sources such as desalination and reclaimed water (Salgot and Tapias, 2004).

Water demand estimation is required before the initiation of new water projects (Downs *et al.*, 2000). This research focuses in particular on water demand by the tourism industry, which is a major water consumer in many arid coastal regions. A model is developed to provide a tool to calculate water demand. The model provides investors (public and private sectors) in water resources projects with the means to estimate current and future water demand required by existing or new tourism facilities in any demand area and time period, also considering, the possibility of introducing demand management measures and different scenarios that will reduce water consumption in hotel facilities. The model calculates demand from several demand nodes, i.e. hotels and adjacent domestic population.

The research involved collecting and analysing data on Sharm El Sheikh, South Sinai, Egypt. The data collected included descriptive hotel data, hotel occupancy rates, monthly water consumption, etc. These data are used to determine specific water consumption rates to be used in the model.

Water consumption of hotels in Sharm El Sheikh

Potable water consumption

The tourism industry requires huge quantities of water supply, with peak consumption during the peak season and excess capacity in the low season (Chartzoulakis *et al.*, 2001). In Sharm El Sheikh, water consumption may be as high as 500 l/d per bed (Hafez and El Manharawy, 2002).

The major water source is desalinated seawater. Depending on the occupancy rate (which varies from 65% to over 100%), water demand fluctuates over the year in this region. Seven 5-star hotels in Sharm El Sheikh were studied, to collect monthly water consumption, occupancy rates, room prices, etc. Hotel water consumption and descriptive data were obtained from the engineering department of each hotel. Hotels 2-7 belonged to the same management. Data was studied for two years 2003 and 2004.

In hotels 1-7, staff lived in a designated area inside the premises of the hotel. However, this is not a common practice for other hotels in the area. Table 5.1 gives some general information on the studied hotels.

Figure 5.1 shows average daily water consumption for the hotels (not considering occupancy rate). As the number of rooms increases in a hotel, specific water consumption per room decreases due to economies of scale. After reaching a break-even point, water consumption increases again.

Another factor influencing water consumption is the price per room. A certain room rate reflects certain standards, sizes, facilities of a hotel, which influence the water consumption. At the same time, the room rate influences the type of guests that visits the hotel, which may also impact water consumption. Figure 5.2 shows change in average water consumption in relation to room rates.

With cheap room rate (59 Euros/room), water consumption was relatively high and with increasing rates water consumption decreased to reach a minimum at nearly 100 Euros/room, before increasing again. According to hotel management, this behavior is related to type of guest, number of guests per room and to the facilities present in a hotel with a given high or low room rate.

Occupancy Rate

Figure 5.3 shows variation of water consumption for hotel 1 with occupancy rate. As the occupancy rate increases, the average consumption per rooms generally decreases (Trung and Kumar, 2005). This relationship is illustrated in Figure 5.4.

Table 5.1: Some characteristics of the surveyed hotels

	Hotel 1	Hotel 2	Hotel 3	Hotel 4	Hotel 5	Hotel 6	Hotel 7
Source of water	Private desalination co.	Hotel-owned desalination plant					
No. of rooms	351	64	82	126	365	133	627
No. of pools	1	1	1	1	1	1	1
Price €/room	80	171	118	94	76	67	59
Type of rooms	Standard	De luxe	Executive	Plus	Standard	Standard	Economy
No. of Staff[1]	370	67	86	132	383	140	658
Avg. potable water consumption m^3/d[2]	494	281	234	314	476	269	653
Avg. cons $m^3/d/room$	1.41	2.19	1.64	1.67	1.57	1.70	1.71
Avg. non-potable water consumption m^3/d [3]	100	42	54	83	240	88	413
Avg. Occupancy rate[4]	72%	66%	66%	66%	66%	66%	66%

[1] Staff number for hotels 2-7 are assumed as 1.05*no. of rooms (consultants' recommendations).
[2] Including potable water used for irrigation (10-30% of total irrigation consumption).
[3] Non-potable water used for irrigation (70-90% of total irrigation consumption); theoretical ratio between potable to non-potable water consumption used for irrigation is 1:9.
[4] For hotel 1, monthly occupation rates were available for years 2003-2004, for hotels 2-7 only average yearly occupation rate was available.

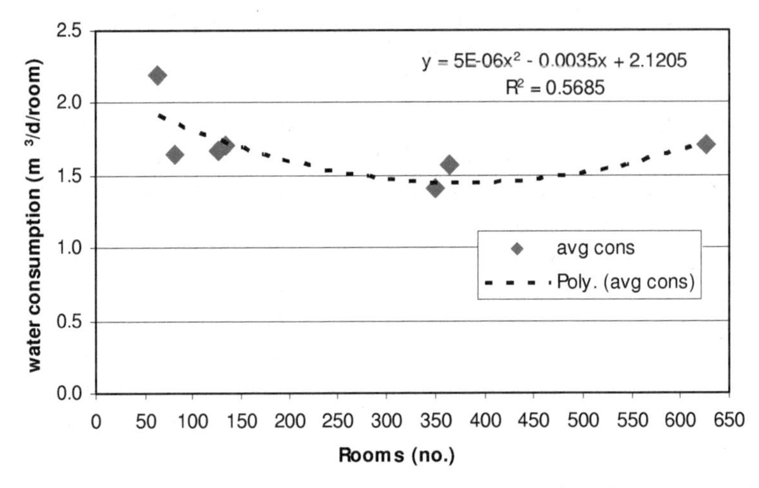

Figure 5.1 Average daily water consumption per room for surveyed hotels

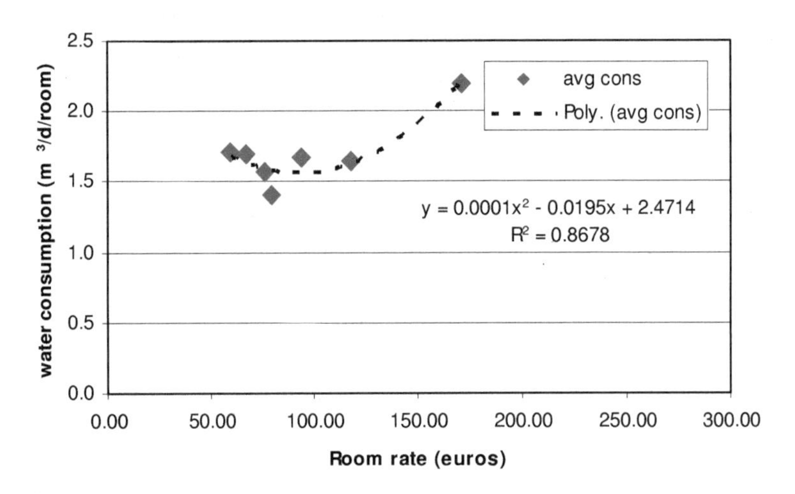

Figure 5.2 Average daily water consumption per room as a function of room rate

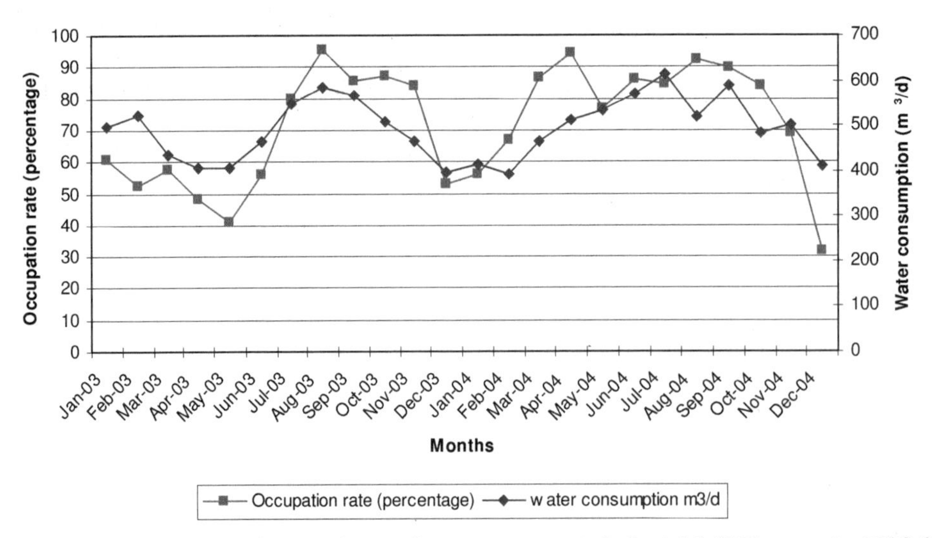

Figure 5.3 Water consumption and occupancy rate in hotel 1 (351 rooms); 2003-2004

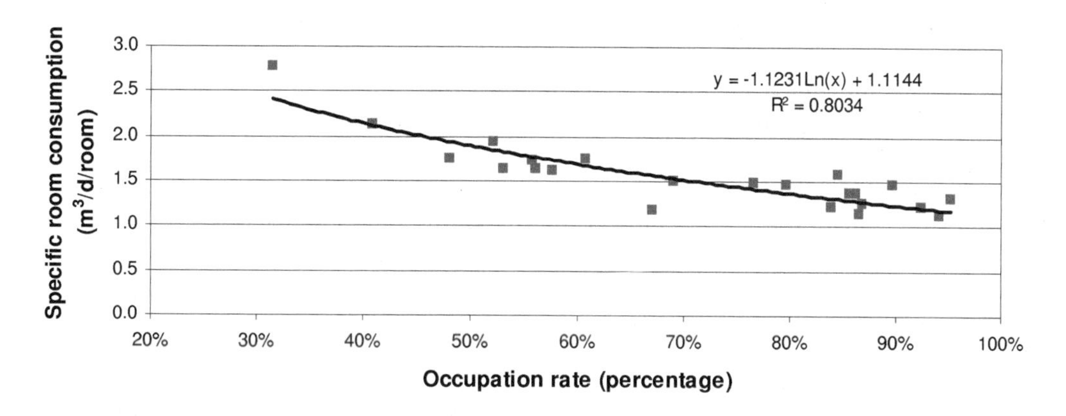

Figure 5.4 Relationship between specific room consumption and occupancy rate; hotel 1

Demand calculation model

The demand calculation model is developed using Excel spreadsheets. The model estimates water demand taking into consideration monthly variations. The model calculates current and future water demand, with and without demand management.

Calculation of current demand

In order to better understand the factors influencing hotels water consumption, a multiple linear regression (MLR) analysis was carried out for the 7 hotels. The following factors were taken into consideration: temperature, occupancy rate and number of rooms. The demand function for hotels is as follows:

$$Q_{dom} = b + d \times T + e \times O \times N_r \qquad \text{Equation 5.1}$$

where Q_{dom} is the water demand of the hotel (m^3/d); *b, d* and *e* are constants; *T* is air temperature expressed as the deviation from the long-term annual mean (°C); *O* is the monthly occupancy rate as a fraction (-); N_r is the number of rooms.

To calculate water demand, the model uses the following equations depending on the details of available data:

$$Q_{dom} = SC_r \times O \times N_r + SC_{sh} \times N_s \qquad \text{Equation 5.2}$$

where SC_r is specific consumption per room including all side activities (m^3/room/d); *O* is monthly occupancy rate; N_r is number of rooms; SC_{sh} is specific consumption of staff housing (and not due to staff activities)(m^3/staff/d); and N_s is number of staff. (SC_r is suggested by the model as a function of the entered occupancy rate).

Or

$$Q_{dom} = SC_r \times O \times N_r + SC_p \times A_p + SC_{gr} \times A_{gr} + SC_{sh} \times N_s + e_f \qquad \text{Equation 5.3}$$

where SC_r is specific consumption per room including only guest-related consumption (m^3/room/d); SC_p is specific consumption for swimming pool (m^3/m^2/d); A_p is the area of swimming pool (m^2); SC_{gr} is specific consumption of irrigated land (m^3/m^2/d); A_{gr} is area of irrigated land (m^2); SC_{sh} is specific consumption of staff housing; N_s is number of staff; and e_f is other fixed water consumption, e.g. boilers, chillers, etc. (m^3/d).

Besides hotels, the model calculates demand from other sources of water demand existing in the studied zone, e.g. local population, agriculture, and industrial demand.

Water demand projection
The following equation is used to project future monthly water demand:

$$D_{dom,j+1} = D_{dom,j} e^{rj} \quad \text{Equation 5.4}$$

where $D_{dom,j+1}$ is future hotel water demand for month j (m^3/d or m^3 per month); $D_{dom,j}$ is hotel water demand in the base month (m^3/d or m^3 per month); r is monthly growth rate in water consumption ($month^{-1}$).

For hotels, the yearly growth rate is assumed to reflect increase in water consumption due to expansion of hotel (construction of new rooms), and/or increase in occupancy rate.

Demand management in model
Demand management in hotels can take the form of awareness programs for visitors and staff, tight maintenance schedules, retrofitting water outlets with water efficient devices, etc. (Schachtschneider, 2000). Generally, efforts are concentrated in guestrooms and include such standard measures as the use of low-flow showerheads, water saving toilets, and linen and towel reuse programs (Gopalakrishnan and Cox, 2003; Meade, 1998).

However the areas where major water conservation has the greatest impact is usually neglected (e.g. kitchens, bars, staff changing rooms, and irrigated landscapes). This was the case in the hotels studied. In order to be able to account for demand management in guestrooms and other facilities, water demand is calculated using the following equation (Froukh, 2001):

$$D_{dom,d} = D_{dom}(1 - C_{eff}/100) \quad \text{Equation 5.5}$$

where $D_{dom,d}$ is hotel water demand applying demand management; D_{dom} is hotel water demand without demand management (m^3/d), as estimated with Equations 5.1, 5.2 and 5.3; C_{eff} is conservation effectiveness expressed as a percentage.

Results

Current Demand
Table 5.2 and Figure 5.5 show multiple linear regression results for the seven hotels. The MLR results show good correlations. However, the MLR did not explain fully the different factors influencing hotels water consumption. Beside occupancy and temperature, a main factor in influencing consumption is “back-of-the house’ activities which are independent of occupancy, climate and irrigation needs (Meade, 1998). The following formula is obtained for the MLR:

$$Q_{dom} = 328.6 + 2.5 \times T + 0.5 \times O \times N_r \quad \text{Equation 5.6}$$

Equation 5.6 implies that:

- Constant b (330 m^3/d) represents water consumption that are more or less fixed and independent of variables included in the MLR, e.g. staff, irrigating landscape, swimming pool, etc.;
- Constant d (2.5 $m^3/$ °C/d) indicates that an increase of 1 °C above average, increases water consumption by 2.5 m^3/d;
- Constant e (0.5 m^3/occupied room/d) indicates that an occupied room consumes on average 0.5 m^3/d.

Table 5.2 Multiple linear regression results for 7 hotels

Unit	Parameter	Hotel 1	Hotel 2	Hotel 3	Hotel 4	Hotel 5	Hotel 6	Hotel 7	Average
m^3/d	b	349	125	112	175	469	188	882	329
°C	d	4.1	0.5	0.8	1.2	3.5	1.3	6.4	2.5
fraction	e	0.6	0.4	0.5	0.5	0.5	0.5	0.5	0.5
no.	N_r *(rooms)*	371	64	82	126	365	133	627	
	r^2	0.61	0.74	0.74	0.74	0.75	0.74	0.74	

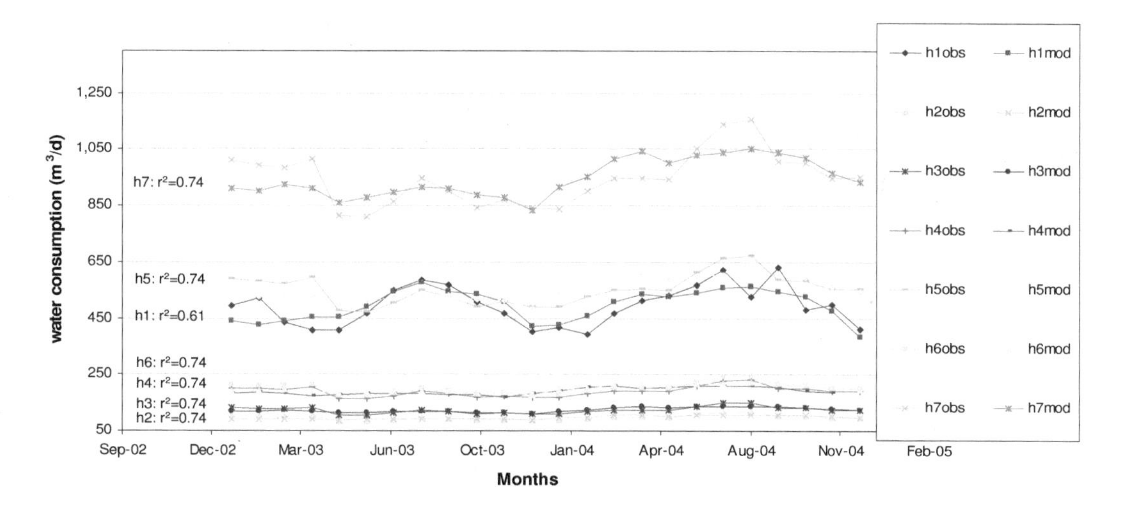

Figure 5.5 Observed and modeled hotels water consumption using MLR

Demand Projection

The user can input a chosen growth rate based on planned hotel expansion or increase in occupancy rate. For studied hotels, an increase in occupancy rate is expected. No hotel expansion was foreseen in the near future. Hotel 1 occupancy rate increased by 14% from year 2003 to 2004, compared to about 30% increase for hotels 2-7. However, through discussion with hotel management, this increase was not typical and average growth rate can be taken as 2-5% yearly.

Demand management

The model calculated water consumption assuming no demand management as one scenario, and considering demand management measures as another. A consideration of demand management options is necessary and these options should run concurrent with the implementation of the supply infrastructure (Dube and van der Zaag, 2003).

Conclusion

Water scarcity is a problem that can constrain the economic development in arid coastal regions. Many of these regions depend primarily on tourism, a sector that does not seem to tolerate water shortage. The nature of development in the Sinai region will put an increasing strain on available water resources and will require the development of new projects.

The development of new water supply projects that can satisfy current and future water demand requires careful calculation of expected water demand taking into consideration previous and future trends. The developed model is a flexible tool which can calculate water demand from various demand nodes. The demand nodes are mainly hotels, adjacent domestic population near hotels and fresh water irrigated landscape.

Water consumption of hotels fluctuates throughout the year. From the case study it was observed that specific water consumption depends on the occupancy rate. The higher the occupancy rate, the lower the specific water consumption per room. The model calculates current and future water demand considering different scenarios of growth and water demand management.

Acknowledgement
Special thanks to Mr. Khaled el Abassy and Mr. Yasser Ibrahim for providing data and information valuable for this research and the outcome of this paper.

6 Optimum contracted-for water supply for hotels in arid coastal regions

Abstract

Hotels in arid coastal areas use mainly desalinated water for their domestic water demands, and treated wastewater for irrigating green areas. Private water companies supply these hotels with their domestic water needs. There is normally a contractual agreement stating a minimum requirement that has to be supplied by the water company and that the hotel management has to pay for regardless of its actual consumption ("contracted-for water supply"). This paper describes a model to determine what value a hotel should choose for its contracted-for water supply in order to minimize total annual water costs. An example from an arid coastal tourism-dominated city is presented: Sharm El Sheikh, Egypt.

Hotels with expected high occupancy rates (74% and above) can contract for more than 80%. On the other hand, hotels with expected lower occupancy rates (60% and less) can contract for less than 70% of the peak daily domestic water demand. With a green area ratio of 40 m^2/room or less, an on-site wastewater treatment plant can satisfy the required irrigation demand for an occupancy rate as low as 42%. Increasing the ratio of green irrigated area to 100 m^2/room does not affect the contracted-for water supply at occupancy rates above 72%; at lower occupancy rates, however, on-site treated wastewater is insufficient for irrigating the green areas. Increasing the green irrigated area to 120 m^2/room increases the need for additional water, either from externally sourced treated wastewater or potable water. The cost of the former is much lower than the latter (0.58 versus 1.52 to 2.14 US$/$m^3$ in the case study area).

Keywords: Tourism; water demand and supply; wastewater reuse; irrigation; water contracts.

This chapter is based on:
Lamei, A., von Münch, E., van der Zaag, P. and Imam, E. (2009b) Optimum contracted-for potable water supply for hotels in arid coastal regions. *Water Sci. & Tech.*, **59**(8), 1541-1550.

Introduction

Tourism has become of prime economic importance for many developing countries (Gossling, 2000). This can cause problems if the tourism is located in an area with water scarcity such as arid coastal regions. Seawater desalination is often the main water source here, and is mainly performed by the private sector under certain contractual arrangements with the tourism sector. The most commonly used contract type is the BOO (Build, own, and operate) contract (Hafez and El Manharawy, 2002). The water company (contractor) will construct and operate the desalination plant, with contracts for water sales specifying a minimum agreed-upon water quantity.

This paper focuses on arid coastal regions, using Sharm El Sheikh (Sharm) in Egypt as an example. In such arid coastal regions it is common practice that hotels buy water from a (private) water company in the vicinity. There may be several private water companies in a city, each supplying water within a certain service area. The service areas are either assigned to them by the local municipality based on concession rights, or are determined by distance and associated transportation costs (the greater the distance of the hotel from the water company, the higher the transportation cost). A contract between the hotel and the water company specifies the amount of water (contracted-for water supply) that has to be bought daily. The hotel management will have to pay for that water whether it is actually consumed or not.

Water in excess of the contracted amount is supplied by the water company only if it is available. Since the water company usually has contracts for the total capacity of its water production plant, no excess water is available in most cases. If the hotel's actual water demand is higher than the contracted-for water supply then hotels may have to buy water from other sources, for example other water companies supplying water by tankers.

Similar contractual arrangements are practiced by water companies in coastal regions in Egypt (Khalil, 2004) and also worldwide in areas where the tourism industry is the dominant water user, e.g. in Greece and Spain (Avlonitis *et al.*, 2002; Gasco, 2004; Khalil, 2004). The user has to buy a minimum amount of water over a certain period of time. Further examples of this practice are the Dhekelia and Larnaca reverse osmosis (RO) desalination plants in Cyprus (WDD, 2007), the Consolidated Water Co. in The Cayman Islands (EDGAR, 2007) and the Hamma RO desalination plant in Algeria (Sadi, 2004).

Hotel managers usually calculate the contracted-for supply as a percentage of their peak daily water demand. For example, in Sharm with a hot arid climate, the peak daily water demand is expected to occur in August where hotels' occupancy rates approach 100%, and the ambient temperature is at its maximum (approximately 40 °C). Typically, the contracted-for water supply for hotels in Sharm is taken to be between 75 to 80% of the peak daily demand (Khaled, 2008; Mohsen, 2007).

If the contracted-for water supply is too high, then the hotel management will spend more money on water than needed. But if the contracted-for water supply is too low, then the hotel management may have excessive costs for buying additional water on an ad-hoc basis, particularly during the peak summer months.

Hotels form a major part of the tourism industry. To enhance profitability, all of the hotel's resources including water should be efficiently managed (Choi *et al.*, 1997). In the literature, several studies have been conducted to optimize the capacity of the water plant (supply node) in order to minimize the cost of water production borne by the water company (Mahmoud *et al.*, 2002; Voivontas *et al.*, 2003). However, according to the authors' knowledge, there have been no studies to optimize the contracted-for water supply (i.e. minimizing cost of water paid by the end user). This paper presents a methodology to optimize the contracted-for water supply for hotels.

Contracted-for water supply optimization method for hotels

Description of contractual agreements between the hotel management and the water company in Sharm

The description below is given for the case of Sharm in Egypt, but as mentioned above, the situation is similar in other tourism-dominated arid coastal regions. A contract is negotiated between a water company and a hotel stating that the water company will supply an agreed upon daily amount (m^3/d). The hotel management has to pay for this agreed amount (contracted-for water supply) even if the hotel's actual consumption is less. The water consumption is measured monthly by water meters at the inlet of the water storage tanks in each hotel and monthly bills are sent to the hotels. The difference between the actual demand and the contracted-for water supply is checked either every 6 or every 12 months. If the amount consumed is less than the agreed upon contracted-for water supply, then no action is taken (it means that the hotel spent more money on water than necessary). If the amount consumed is more than the contracted-for water supply, the hotel management has to settle the difference (Saleh, 2008).

The contracted-for water supply is fixed for the duration of the contract. The contract between the water company and the hotel management has a long duration, typically 15 or 20 years. If the hotel wants to increase the contracted-for water supply then it has to notify the water company 9 months in advance in order for the water company to increase its capacity.

Water companies often have difficulties to meet the contracted-for supplies from hotels. The reasons are:

- Hotels expand their room numbers without prior notification to the water company;
- Hotels do not take into consideration that the water company is supplying the contracted-for demand over a period of 24 hours and that instantaneous demand could not be met;
- Some clients consume more than the contracted-for water supply putting extra pressure on the network and resulting in shortage for other clients who are still within their water budget. In these cases, the water company has the obligation to supply the shortage usually by water tankers at their own expense. In some cases, the water company can limit the supply to such clients by shutting down the pipeline for certain excessive consumers.

If the water company cannot supply enough water, and the hotel's demand exceeds the contract-for amount, then the hotel usually has to buy the extra from a source other than the water company at a much higher price (Table 6.1). The hotels usually have storage tanks for only one or two days to balance daily variations in demand.

Table 6.1 Unit prices of different types of water in the case study area of Sharm (Khaled, 2008)

	Potable water (P_P)	Excess potable water (P_{EP}) (avg. of 3 water companies)	Treated wastewater on-site (P_{TWW})[a]	Excess wastewater[b] (P_{WW})
Selling prices (US$/m^3)	1.52	2.14	0.33	0.58

[a]Cost incurred by hotel management for treating wastewater.
[b]Selling price of treated wastewater charged by Al Montaza Co.

Contracted-for water supply model

Overview
Prediction of a future hotel's water consumption is complicated as it involves analysis of many factors including occupancy rate, hotel outdoor area (including green irrigated area and pool), type of laundry, facilities, number of restaurants, type of cuisine, class of hotel, type of water used for irrigation, and presence of staff quarters (Deng and Burnett, 2000).

A model to optimize the contracted-for water supply for hotels is described in this section. The model calculates what the contracted-for water supply of a given hotel should be, based on the hotel's characteristics on the one hand (such as occupancy rate, size of green irrigated area) and the unit prices of the different types of water on the other hand.

The contracted-for water supply model is divided into three sub-models: water demand, water supply and cost models:

- The water demand model calculates the hotel's water demand of different users: guests, pool, staff, and irrigated green area. It should be pointed out that all water uses, except irrigation, requires water of potable quality. Irrigation demand can be met with treated wastewater.
- The water supply model calculates the quantity of contracted-for water which is needed from the water company and the amount of extra water needed (in excess of the contracted-for water supply). Excess water can be supplied by the water company itself (depending on availability) and/or from private individual trucks (at higher water prices than from the water company).
- The cost model calculates the cost of water to hotels in periods of high and low demand: cost of contracted-for water, cost of excess water if needed, cost of treating wastewater (ww) on-site, and cost of buying treated wastewater from an external source.

The paper uses Sharm in Egypt to demonstrate the methodology. Table 6.2 lists the input parameters of the model, built-in values and expected output. Figure 6.1 is a flow chart describing the contracted-for water supply optimization model for hotels.

Table 6.2 List of input, built-in values and output of contracted-for water supply model for hotels

Type of variable	Symbol	Unit	Description
Input	A_{gr}	m^2	Area of irrigated land
	A_p	m^2	Area of swimming pool
	f_{un}	%	Factor for unaccounted-for water
	f_{wdm}	%	Factor for water demand management level
	N_r	-	Number of rooms
	O	%	Average monthly occupancy rate
Built-in values	f_{dw}	%	Factor indicating the percentage of domestic water demand ending up as sewage
	f_g	-	Fraction for number of guests per room
	f_s	-	Fraction for number of staff per room
	SC_g	$m^3/c/d$	Specific water consumption of guests
	SC_{gr}	$m^3/m^2/d$	Specific water consumption of irrigated land
	SC_p	$m^3/m^2/d$	Specific water consumption of swimming pool
	SC_s	$m^3/c/d$	Specific water consumption of staff
	SC_{sh}	$m^3/c/d$	Specific water consumption of staff housing
Output	C_T	US$/yr	Minimum yearly cost of domestic and irrigation water
	Q_c	m^3/d	Daily contracted-for water supply
	Q_{dom}	m^3/d	Daily domestic water demand of a hotel
	Q_I	m^3/d	Daily water demand of irrigation
	Q_{peak}	m^3/d	Peak water demand of a hotel
	Q_{ww}	m^3/d	Daily wastewater flow rate

Water demand model

The water demand of a hotel consists of two parts: domestic water demand and irrigation water demand. In this section equations are presented to estimate both types of demand.

Domestic water demand

The fresh water demand of a hotel is a function of the number of hotel rooms, probable occupancy rate, number of staff working and/or living in a hotel, presence and size of swimming pool, kitchen, laundry, etc. A swimming pool water can be supplied from fresh or saline water. The domestic water demand does not include water for the irrigation of green garden areas since lower quality water can be used for that purpose. The following equation calculates the daily domestic water demand for hotels:

$$Q_{dom,j} = (SC_g \times f_g \times O_j \times N_r + SC_{p,j} \times A_p + N_s \times (SC_s + SC_{sh})) \times f_{un} \times f_{wdm}$$

Equation 6.1

where $Q_{dom,j}$ is the domestic water demand in month j (m^3/d), SC_g is the specific consumption per guest; f_g is a fraction for number of guests per room; O_j is the average daily occupancy rate in month j (%); N_r is the number of rooms in a hotel; $SC_{p,j}$ is the specific consumption for the swimming pool in month j ($m^3/m^2/d$); A_p is the total pool area (m^2); SC_s is the specific consumption of staff during working hours ($m^3/cap/d$); N_s is the number of staff; SC_{sh} is the specific consumption of staff housing ($m^3/cap/d$); f_{un} is a factor to estimate unaccounted water in a hotel e.g. cleaning pathways, leakage, etc.; and f_{wdm} is a factor to account for the level of water demand management practiced in the hotel. Example values for these parameters are provided in Table 6.3.

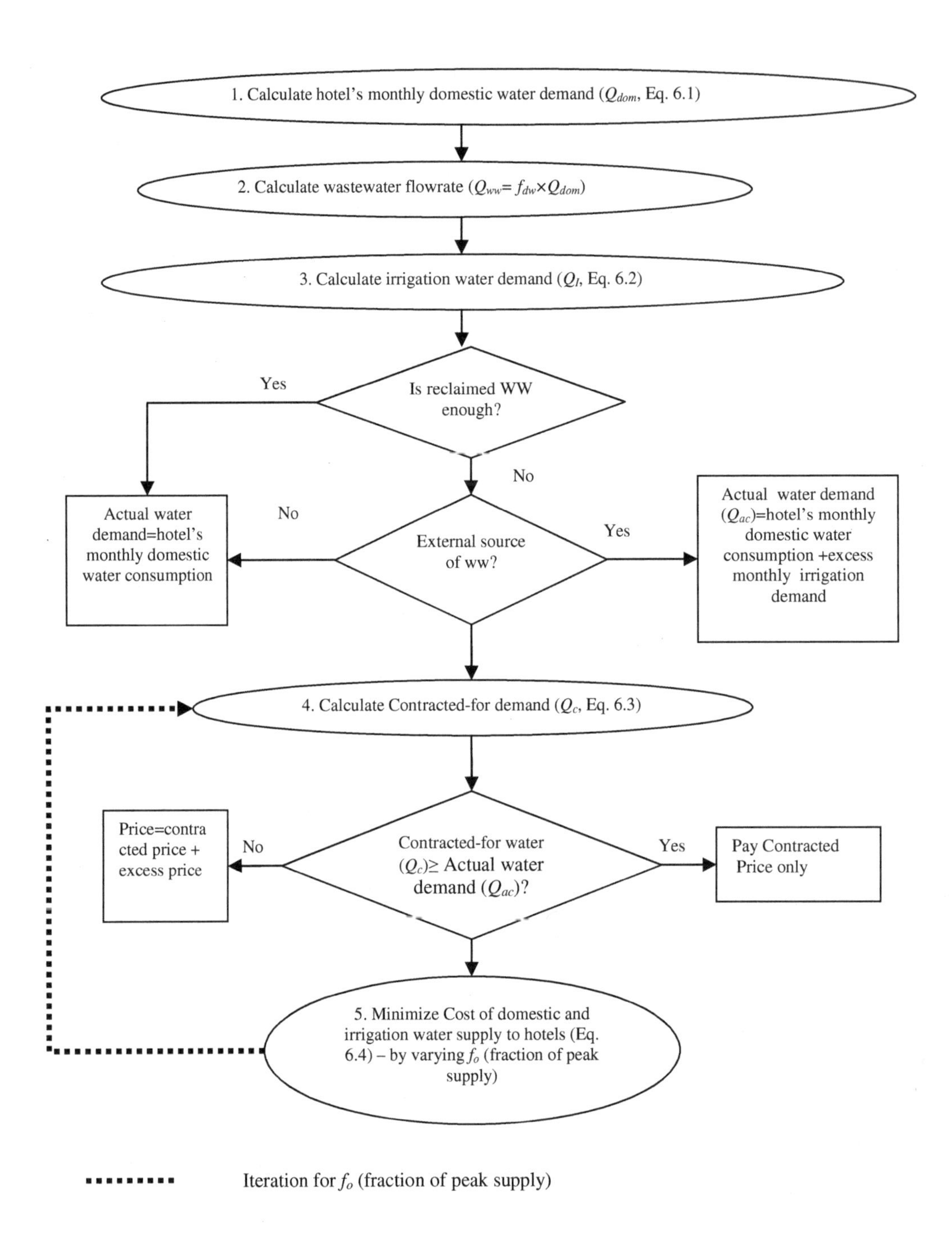

Figure 6.1 Flow chart of contracted-for water supply model for hotels

Irrigation water demand

The irrigation water demand is a function of the irrigated green area. The green area could vary from a golf course, large green area or small green area. There are no common standards for the ratio of green area to the number of rooms. Usually 3-star hotels are expected to have smaller green areas than 4 or 5-star hotels. The irrigation water demand is calculated as follows

$$Q_{I,j} = SC_{gr,j} \times A_{gr} \qquad \text{Equation 6.2}$$

where $Q_{I,j}$ is the irrigation water demand in month j (m^3/d); $SC_{gr,j}$ is the specific consumption of the green area as a function of temperature in month j ($m^3/m^2/d$); A_{gr} is the area of irrigated land (m^2). Example values for these parameters are provided in Table 6.3.

Supply model

The water supply consists of two parts, namely potable water supplied from the water company, and irrigation water from treated wastewater. Figure 6.2 shows the water supply scheme for domestic and irrigation water demand for a typical hotel in an arid coastal region. If there is an external source selling treated wastewater then the irrigation water needed over and above the capacity of the hotel's own wastewater treatment plant (WWTP) can be sourced from that external seller of treated wastewater. If not, excess demand should be met from fresh water supplied to the hotel from the water company.

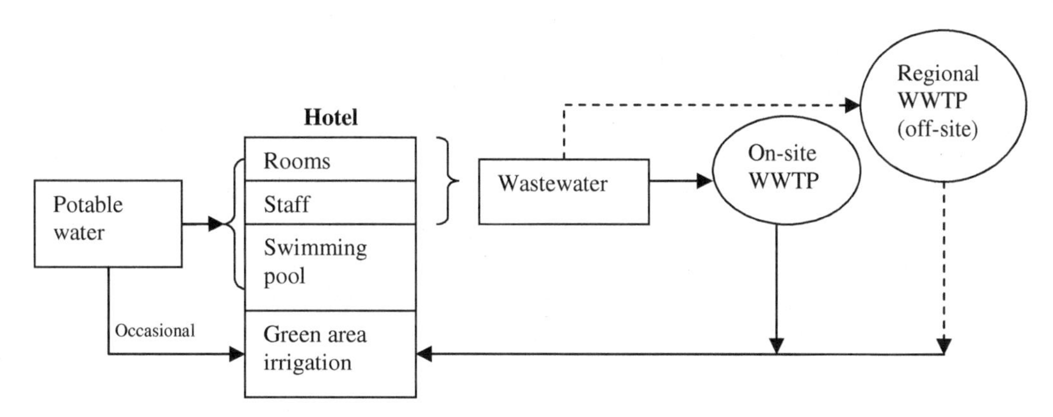

Figure 6.2 Domestic and irrigation water supply scheme for hotels in an arid region such as Sharm, Egypt

Potable water supply

The contracted-for supply would be calculated as follows:

$$Q_{c,j} = f_o \times Q_{peak,j} \qquad \text{Equation 6.3}$$

where $Q_{c,j}$ is the contracted-for water supply in month j (m^3/d); $Q_{peak,j}$ is the peak domestic water demand in month j (m^3/d) (calculated using Equation 6.1 with 100% occupancy rate and highest expected temperature; in this case the month of August). Q_{peak} does not include the demand for irrigation water; f_o is a fraction of the peak demand.

The actual potable water demand can be higher than contracted-for water supply due to seasonal variations in occupancy rate. In addition, excess irrigation demand which is not met by an external source selling reclaimed wastewater will be met by potable water.

Irrigation water supply: wastewater recycling and reuse
A proportion of used domestic water results in wastewater production. Typically, this proportion is about 80% of used potable water (Khaled, 2008; Mohsen, 2007). Wastewater is treated and reused for irrigation of landscape within hotels resulting in savings in potable water (Hamoda, 2004).

Cost model
The objective function is to minimize yearly cost of potable and treated wastewater for the hotel as well as to satisfy the peak expected demand of the hotel. The equation can be expressed as follows:

Minimize

$$C_{T,y} = \sum_{j=1}^{12} P_P \times Q_{c,j} + P_{EP} \times \max(Q_{dom,j} - Q_{c,j}, 0) + P_{TWW} \times f_{dw} \times Q_{dom,j} +$$

$$\min(P_{WW}, P_{EP}) \times \max(Q_{I,j} - f_{dw} \times Q_{dom,j}, 0)$$

Decision variable: f_o Equation 6.4

where $C_{T,y}$ is the total cost for domestic and irrigation water supply in year y (U$/yr); P_P is the selling price of potable water (US$/m^3); $Q_{c,j}$ (see Equation 6.3) is the contracted-for water supply in month j (m^3/month; daily values multiplied by 30 days); P_{EP} is the selling price of potable water in excess of contracted-for water supply (US$/m^3); $Q_{dom,j}$ (see Equation 6.1) is the domestic water demand in month j (m^3/month; daily values multiplied by 30 days); P_{TWW} is the unit cost of treating wastewater within the premises of the hotel (US$/m^3); f_{dw} is a factor indicating the percentage of domestic water demand ending up as sewage; P_{WW} is the selling price of treated wastewater bought from an external source (US$/m^3); $Q_{I,j}$ (see Equation 6.2) is the irrigation water demand in month j (m^3/month; daily values multiplied by 30 days); f_o is a fraction of the peak demand.

Example model results

Case study area
An example from the city of Sharm in Egypt is presented in this section. There are three main private water companies operating in Sharm, each with a service area (concession rights): South Sinai Water Co., Al Nabq Water Co., and Sinai Environmental Services. As each water company usually has contracts for the total water produced from its plant, there is rarely any excess water and, if needed, hotels will then only have the option of buying from water trucks at a higher price (water trucks operators either buy water themselves from other water companies and then resell it at a higher price, or simply act as a transporter charging a fee for transportation).

At the moment, there are no major private wastewater treatment companies selling treated wastewater back to hotels except for one company (Al Montaza) established in late 1995 which has a network connected to a group of 15 hotels. The company collects the hotels' wastewater for free and sells it back after treatment to the same hotels at a fee. Most hotels have on-site wastewater treatment plants; for some small hotels, wastewater is discharged to a government-owned wastewater treatment plant where it is treated (primary treatment) and piped to irrigate 100 acres of tree plantations owned by the local municipality.

Unit costs used in the model calculation are shown in Table 6.1. Costs are shown in US$ using an exchange rate of 1 US$=5.6 Egyptian pounds. Table 6.3 shows an example of model calculations based on local data. These numbers can be adjusted based on location.

The results showed that contracted-for water supply is a function of several parameters: occupancy rate, size of green irrigated area, and source of irrigation water.

Occupancy rate

Hotels with expected high occupancy rates (74% and above) can contract for more than 80%. If they contract for less, this would mean higher costs imposed on them in terms of having to buy excess water at a higher price in periods of high demand. On the other hand, hotels with expected lower occupancy rates (60% and less) can contract for less than 70% of the peak daily domestic water demand. Figure 6.3 shows the effect of varying the occupancy rate on the optimal fraction of peak water demand to be contracted, as well as the total cost of potable water supply.

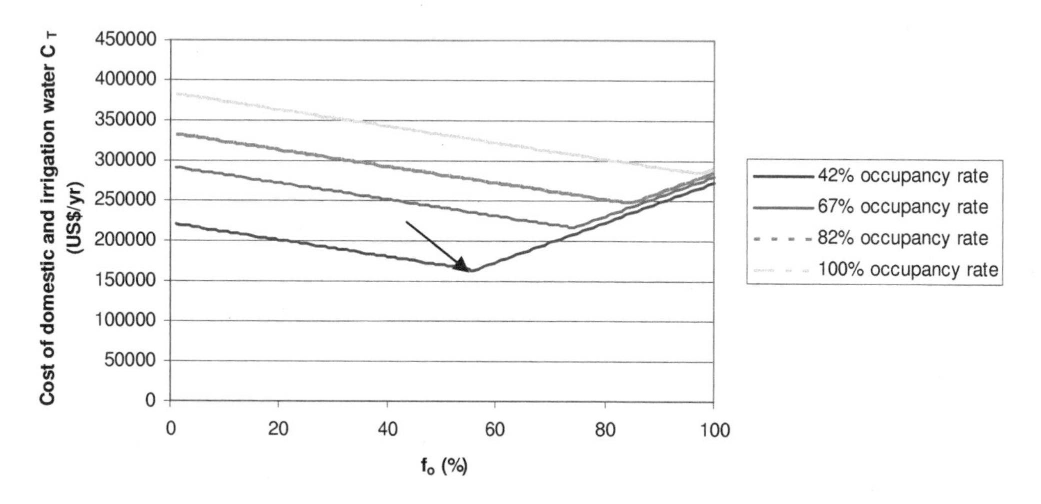

Figure 6.3 Total water cost C_T versus f_o (fraction of peak demand) for different average yearly occupancy rates O; (Arrow points to C_T in Table 6.3; model parameters are shown in Table 6.3 unless otherwise indicated)

Table 6.3 Example of water demand calculations to determine optimized f_o value for a hotel in Sharm

Model Input	Symbol	Unit	Example values	Comments
Area of pool	A_p	m^2	1,600	
Green area	A_{gr}	m^2	14,500	Ratio of green area to number of rooms is to $41 m^2$/room
Fraction for unaccounted-for water	f_{un}	-	0	
Fraction for water demand management level	f_{wdm}	-	0	
No. of rooms	N_r	-	351	
Average yearly occupancy rate	O	%	42	Average monthly occupancy rate ranges from 26% in January to 75% in August
Staff housing within premises		-	yes	Hotels can have staff housing within premises or outside
Unit prices of different types of water		US$/$m^3$	Table 6.1	
Model built-in parameters				Use same values unless other information is available
Factor indicating the percentage of domestic water demand ending up as sewage	f_{dw}	%	80	(Khaled, 2008; Mohsen, 2007)
Fraction for number of guests per room	f_g	-	2.3	Average from nine surveyed hotels in Sharm (Chapter 5)
Optimized fraction of peak demand	f_o	%	100	Value of f_o before iteration
Fraction for number of staff per room	fs	-	1.05	(Chapter 5)
Staff number	N_s	-	367	Ratio of staff to rooms (see previous line) multiplied by no. of rooms in a hotel
Specific water consumption of guests	SC_g	m^3/cap/d	0.4	Average from nine surveyed hotels in Sharm (Chapter 5)
Specific water consumption of irrigated land	SC_{gr}	$m^3/m^2/d$	0.01	From an empirical equation of average monthly temperature[a]
Specific water consumption of swimming pool – if using fresh, not saline water	SC_p	$m^3/m^2/d$	0.01	From an empirical equation of average monthly temperature (Chapter 5)
Specific water consumption of Staff housing	SC_{sh}	m^3/cap/d	0.25	Average from nine surveyed hotels in Sharm (Chapter 5)
Specific water consumption of staff	SC_s	m^3/cap/d	0.03	Average from nine surveyed hotels in Sharm (Chapter 5)
Average yearly temperature	T	°C	29	Average monthly temperature varies from

				18 in January to 42°C in August
Model output				
Optimised fraction of peak demand	f_o	%	56	(After iteration) based on objective function to minimize cost of domestic and irrigation water (C_T)
Minimum yearly cost of domestic and irrigation water	C_T	US$/yr	164,000	Equation 6.4
Domestic peak demand	Q_{peak}	m^3/d	453	From Equation 6.1 using temperature of 42°C (August temp) and 100% occupancy rate
Daily contracted-for water supply	Q_c	m^3/d	253	From Equation 6.3 with Q_{peak} equal to 453 (see previous line)
Average daily water demand of irrigation	Q_I	m^3/d	116	Equation 6.2
Average daily wastewater flowrate	Q_{ww}	m^3/d	204	$Q_{ww} = f_{dw} \times Q_{dom}$
Potable water needed for irrigation		m^3/d	0	Wastewater inflow is more than irrigation daily water consumption ($Q_{ww}>Q_I$)

[a]Equation for calculation of green irrigated area specific consumption SC_{gr}: 0.0047+0.00112T; where SC_{gr} is in m^3/m^2/d and T is temperature in °C (Data collected from 9 hotels in Sharm El Sheikh)

Size of green area and source of irrigation water

The ratio of green area to number of rooms has an impact on required water. Table 6.4 shows the ratio of green area to number of rooms in nine surveyed hotels in Sharm, Egypt. All hotels have on-site wastewater treatment plants. Figure 6.4 shows results of varying occupancy rate and green area on the contracted-for water supply. The optimum percentage for the contracted water supply is the percentage resulting in lowest water cost to hotels.

Table 6.4 Ratio of green area to number of rooms in surveyed hotels

	Hotel 1	**Hotel 2**	**Hotel 3**	**Hotel 4**	**Hotel 5**	**Hotel 6**	**Hotel 7**	**Hotel 8**	**Hotel 9**
Classification	5-star	5-star	5-star	5-star	5-star	5-star	5-star	5-star	3-star
No. of rooms	351	396	64	82	126	365	133	627	107
Green area (m^2)	14,500	57,000	4,218	5,405	8,305	24,058	8,766	41,327	2,400
Green area/no. of rooms (m^2/room)	41	143	65	65	65	65	65	65	22

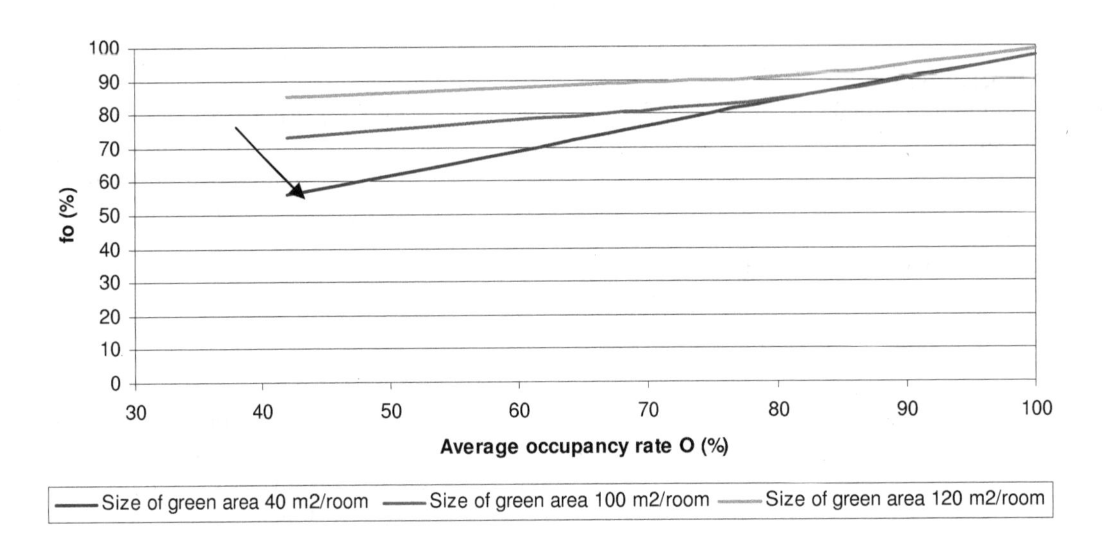

Figure 6.4 Fraction of peak demand for contracted-for water supply f_o (%) versus average yearly occupancy rates O for different sizes of green area; (Arrow points to f_o in Table 6.3; model parameters are shown in Table 6.3 unless otherwise indicated)

With a green area ratio of 40 m^2/room or less, an on-site wastewater treatment plant can satisfy the required irrigation demand for an occupancy rate as low as 42%. Increasing the ratio of green irrigated area to 100 m^2/room does not affect the contracted-for water supply at occupancy rates above 72%; at lower occupancy rates, however, on-site treated wastewater is insufficient for irrigating the green areas. Increasing the green irrigated area to 120 m^2/room increases the need for additional water, either from externally sourced treated wastewater or potable water. Obviously the cost of the former is much lower than the latter (0.58 versus 1.52 to 2.14 US$/$m^3$ in the case study area). However, not only the availability and cost determine which type of water will be used, as also concerns on the quality of treated wastewater may arise.

Instead of using fresh water for irrigation to compensate for shortage in treated wastewater, hotels can buy treated wastewater from an external supplier (at a cheaper cost than fresh water). For a hotel with large green area (> 40 m^2/room), buying external treated wastewater reduced the overall water cost and the amount of contracted-for water supply. Figure 6.5 shows the impact of using treated wastewater for excess irrigation (instead of fresh desalinated water) on the total cost of domestic and irrigation water supply (C_T) for a hotel.

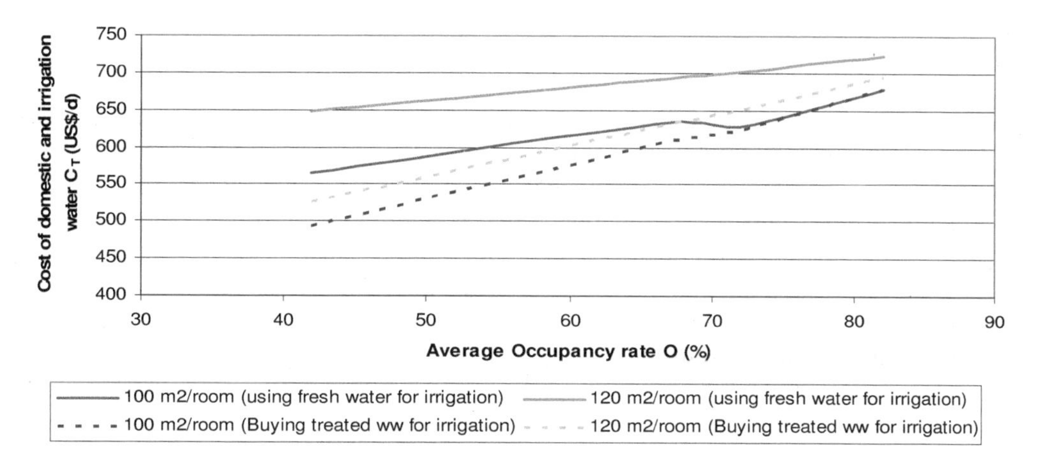

Figure 6.5 Effect of using an external source of treated wastewater on the daily cost of domestic and irrigation water C_T

Conclusions

Hotels buy their water from a water company based on a contract which stipulates a minimum daily amount of water (contracted-for water supply). Water in excess of that amount can be supplied by the water company if available. As the water company usually has contracts for the total water produced from its plant, there is rarely any excess water and hotels will then only have the option of buying from water trucks at a higher price.

Hotel managers have to carefully analyze their hotel's water requirements in order to determine which percentage of the hotel's peak water demand should be used in the contract ("contracted-for water supply") in order to reduce water costs and avoid risk of water shortage. Typically, hotel managers opt for 75-80% of their peak daily domestic water demand based on experience.

This paper presents a methodology to optimize the contracted-for supply for hotels. It analyzes what the contracted-for water supply of a given hotel should be, based on the hotel's characteristics on one hand (such as occupancy rate, size of green irrigated area) and the unit prices of different types of water on the other hand.

The results showed that contracted-for water supply is a function of several parameters including occupancy rate, size of green area and type of water used for irrigation. Hotels with expected high occupancy rates (74% and above) can contract for more than 80%. If they contract for less, they would incur higher costs as they have to buy excess water at a higher price in periods of high demand. On the other hand, hotels with expected lower occupancy rates (60% and less) can contract for less than 70% of the peak daily domestic water demand.

The ratio of green area to number of rooms has an impact on required water. With a green area ratio of 40 m^2/room or less, an on-site wastewater treatment plant can satisfy the required irrigation demand for an occupancy rate as low as 42%. Increasing the ratio of green irrigated area to 100 m^2/room does not affect the contracted-for water supply at occupancy rates above 72%; at lower occupancy rates, however, hotel's treated wastewater is insufficient for irrigating the green areas. Increasing the green irrigated area to 120 m^2/room increases the need for additional water, either from externally sourced treated wastewater or potable water. Obviously the cost of the former is much lower than the latter (0.58 versus 1.52 to 2.14 US$/m^3 in the case study area).

However, not only the availability and cost determine which type of water will be used, as also concerns on the quality of treated wastewater may arise. Prediction of a future hotel's water consumption is complicated as it involves analysis of many factors including occupancy rate, gross compound area (including green irrigated area and pool), type of laundry, facilities, number of restaurants, type of cuisine, class of hotel, type of water used for irrigation, and presence of staff quarters (Deng and Burnett, 2000). Evaluation of the hotel water consumption can help in better projecting future water needs and required water resources.

7 Dynamic programming of capacity expansion of reverse osmosis desalination plant case study: Sharm El Sheikh, Egypt

Abstract

Hotels in arid coastal areas use mainly desalinated water for their domestic water demand, and treated wastewater for irrigating green areas. Private water companies supply these hotels with their domestic water needs. There is normally a contractual agreement stating a minimum requirement that has to be supplied by the water company and that the hotel management has to pay for regardless of its actual consumption ("contracted-for water supply"). With a reverse osmosis (RO) desalination plant designed to satisfy only the contracted-for water supply, the water company would be missing out on potential benefits that could have been obtained selling water in periods of high demands. On the other hand, sizing the RO desalination plant to produce water to satisfy peak demand means incurring additional costs as well as having the plant partially idle during periods of average or low demand.

A model was developed using Excel macros to perform dynamic optimization with the objective function to maximize the present value of total benefits over the lifetime of the RO desalination plant. The aim of the dynamic optimization is to solve for capacity expansion. The model can be used to test different scenarios to capture time-variant tourism demand and price uncertainties on investment decisions.

Unit production cost of RO desalination plants varies according to the degree of operation of the plant. This fact has to be taken into consideration when calculating the costs of RO desalination and when deciding on plant capacity in order to maximize the total net benefits.

Keywords: Dynamic optimization; modeling; water demand; capacity; reverse osmosis; desalination.

This chapter is based on:
Lamei, A., Tilmant, A., van der Zaag, P. and Imam, E. (2009c) Dynamic programming of capacity expansion of reverse osmosis desalination plant Case study: Sharm El Sheikh, Egypt. *Water Sci. & Tech.: Water Supply,* **9**(3), 233-246.

Introduction

Water treatment facilities are usually designed so that the plant satisfies demand for long design horizons. Reverse osmosis (RO) plants are special and should be looked at differently from traditional water treatment facilities.

An RO desalination plant is costly but can be expanded easier than a conventional water treatment plant; therefore, an RO plant can be expanded in several timely stages. The major component of the RO plant is the mechanical equipment (about 67%) including high pressure pumps and membranes whereas civil works constitute about 23% of the total capital cost of the plant (Sommariva, 2004) . This is unlike the conventional water treatment facilities where civil work constitutes the major component of the capital cost.

The capacity of the water treatment plant is usually designed to satisfy the maximum daily demand. Though installing larger capacity equipment can be cheaper due to economies of scale, the mechanical equipment in an RO plant can not be left idle without operation for long durations. The membranes in most of the cases are not preserved correctly, subjecting them either to damage or bacterial re-growth on its surface and may be fouled making it unsuitable for use. Also, unnecessary costs (e.g. unused capital and extensive maintenance at start up) would be incurred due to idle equipment during periods with less than peak demand, thus affecting the economics and the efficiency of the plant. On the other hand, under-sizing the RO plant means loss of potential revenue and a smaller market share.

In Egypt and also worldwide, private RO desalination plants are generally owned by water companies through a BOO (Build, Own, Operate) agreement (Avlonitis *et al.*, 2002; EDGAR, 2007; Gasco, 2004; Hafez and El Manharawy, 2002; Khalil, 2004; Sadi, 2004; WDD, 2007). The water company (contractor) will construct and operate the desalination plant, with contracts for water sales. The main client of private water companies (which is the focus of this paper) is hotels. Other users (homes and restaurants) are supplied by government-owned desalination plants which supply water based on a fixed unit price and with no contractual limitation on the quantity of water supply.

Individual contracts are negotiated and signed between the private water company and each client (a hotel in this case) stating that the water company will supply an agreed-upon daily/monthly amount. The hotel management has to pay for this agreed amount (contracted-for water supply) even if the hotel's actual consumption is less. The duration of the contract can be up to 20 years (Chapter 6). In addition of supplying water to satisfy the total amount stipulated in the signed contracts with the concerned hotels, there is an excess demand arising from the difference between contracted-for supply and average demand, expected actual demand and spike holidays (i.e. public holidays).

A water company has to make a decision on the economic RO capacity it should install along the lifetime of the project in order to maximize its net benefit. The installed RO capacity has to satisfy the signed contracts and a portion of the excess demand. An understanding of the tourism market is likely to aid the water company to estimate future water demand and anticipate future contracts. Unit production cost of an RO desalination plant varies according to the degree of operation of the plant. This fact also has to be taken into consideration when calculating the costs of RO desalination and when deciding on the plant capacity in order to maximize the total net benefits.
According to the authors' knowledge, optimization of multistage capacity expansion of reverse osmosis (RO) desalination plants has not yet been investigated. Several studies

for optimization of multistage capacity expansion of conventional water treatment systems have been published (Dandy *et al.*, 1984; Hinomoto, 1972; Loucks and Van Beek, 2005) using different techniques including dynamic programming.

Dynamic programming is used here to optimize capacity expansion (staging) of RO plants. The model is divided into sub-models: demand prediction, supply and economic models. The demand model calculates the hotel potable and non-potable water demand from different users: guests, pool, staff and irrigated green area. The supply model calculates the quantity of contracted water which is needed from the water company and any extra water (in excess of contracted water). Excess water can be supplied by the water company itself (depending on availability) and/or purchased from an external source (another water company) at a higher price and delivered (bearing extra costs for truck transportation) to the contracted user. The economic model calculates the cost of supplying water to hotels (capital, O&M and production costs) and expected revenue in periods of high and low demand. This study focuses on tourism dominated arid coastal regions, using Sharm El Sheikh (Sharm) in South Sinai, Egypt, as an example.

RO plant capacity expansion optimization

The water company has to satisfy all signed contracts with users. In order to maximize the economic benefits and ensure sufficient water production, the water company can choose to install additional capacity in excess of that required to satisfy contracts for the purpose of satisfying excess demand arising from the difference between contracted-for supply and average demand, expected actual demand and spike holidays.

In order to decide on the RO plant capacity and design a scheme for expansion, an optimization model is developed with the objective function to maximize the present value of total net benefits accrued from capacity expansion while meeting water company's contractual obligations, over the planning horizon of the project.

In this model, the planning horizon of the project can be up to 10 years. Usually RO plants have a lifetime of 10 years (for mechanical equipment) and 30 years for infrastructure. Dynamic programming was done for a period of 5 years. The reason for choosing 5 years is that data available from the RO desalination plants in the case study area were only available for 5 years (2004 to 2008). The following is the objective function to be maximized:

Maximize Net-Benefit

$$B = \sum_{y=1}^{Y} (R_y - C_y) \qquad \text{Equation 7.1}$$

where B is the present value of the net benefit over the planning horizon of the project (US$); R_y is the present value of expected revenue from water sales in year y (US$/yr); and C_y is the present value of the cost of water production in year y (US$/yr); and Y is the planning horizon of the project (yr).

The investment decisions for the model are discrete. The RO plant capacity is increased in an order of 50 m^3/d. The increase can be achieved by adding extra stages or changing the operating pressure of the pumps (Girgis, 2008). The forward-moving dynamic programming algorithm (but backward looking) is used in this problem, i.e. calculating the best value of the objective function that could be obtained from all past decisions leading to that state or node. Figure 7.1 illustrates the idea of dynamic programming showing different paths for capacity additions along the duration of the project.

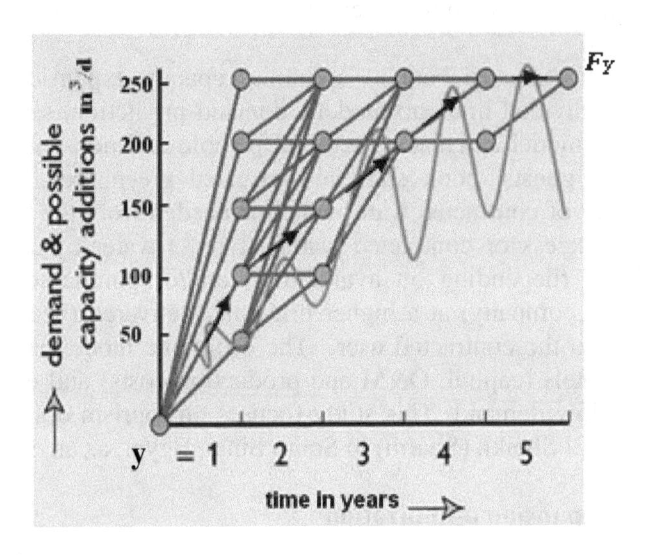

Figure 7.1 Network of discrete capacity-expansion decisions (red links x_y) that meet time-variant projected demand (green line), blue nodes represent discrete states (s_{y+1}) or capacity expansion, F_Y is the maximum present value of the total net benefits, arrows on the red lines indicate the set of best decisions x_y for all stages y that yielded the maximum F_Y (adapted from (Loucks and Van Beek, 2005))

At the beginning of the first year $y = 1$, the accumulated maximum net benefit $f_0(s_1)$ is 0.

For year $y = 1$,
s_2 will range from D_1 to the maximum demand D_Y,
$f_1(s_2) = \max \{R_1(s_1,x_1)-C_1(s_1,x_1)\}$
$x_1 = s_2$ and $s_1 = 0$ Equation 7.2

where s_2 is the state (RO capacity) at end of year 1; D_1 is demand at year 1 (equivalent to contracted-for water supply or basic demand); D_Y is maximum demand reached at year 5 summing up both contracted-for supply and extra potable water demand; R_1 is revenue in year 1; C_1 is unit production cost incurred in year 1; x_1 is capacity expansion in year 1; s_1 is the state (RO capacity) at the beginning of the project.

For all stages between the first and last years,
$f_y(s_{y+1}) = \max\{(R_y(s_y,x_y)-C_y(s_y,x_y)+f_{y-1}(s_y)\}$ over all discrete
x_y between 0 and $s_{y+1}-D_{y-1}$
where $s_y = s_{y+1}-x_y$ Equation 7.3

For the last stage $y = Y$ and for the final discrete state $s_{Y+1} = D_Y$,
$F_Y(s_{Y+1}) = \max \{ R_Y(s_Y,x_Y)-C_Y(s_Y,x_Y)+f_{Y-1}(s_Y)\}$over all discrete
x_Y between 0 and D_Y-D_{Y-1}
where $s_Y = s_{Y+1}-x_Y$ Equation 7.4

In order to identify the sequence of capacity expansion decisions that results in the maximum net benefit (F_Y), backtracking is required to collect the set of best decisions x_y

for all stages y. The solution to Equations 7.1 through 7.4 is presented in the section on the application of the dynamic optimization model on the case study area.

Water demand/supply model

For optimum capacity expansion, current and future water demand have to be calculated and the factors affecting the water consumption of the targeted population have to be analyzed. In this study, the targeted population is tourists. Touristic water demand is a function of several factors including time-variant occupancy rates, number of hotel rooms, specific consumption of guests and other related-services (e.g. pool and staff).

Tourism growth rate follows a different pattern from normal population growth, while the latter usually follows an exponential growth curve, tourism growth rate is affected by external factors including economics, marketing and security. Tourism growth over a period of time is referred to as tourism life cycle.

Tourism life cycle

In 1980, Butler was the first to introduce the theory of a tourism life cycle model (Alvarea and Lourenco, 2005). Recently, his model has been adapted by others. In Spain, the tourism cycle model was divided into several stages (Figure 7.2) as follows:

- A first stage called **discovery**, in which tourism begins to implant itself timidly in a territory for the first time.
- A second stage known as **launch**, in which the tourism phenomenon grows spectacularly and very quickly. There is a change from minority practices to others of general scope, characterized by an enormous quantitative increase of both demand and supply, following a pattern of discontinuous growth.
- A third stage of **stagnation**, in which saturation is reached: the quality of the offer falls, demand levels off, and the environmental degradation of the tourist destination begins to be obvious and worrying.
- A fourth stage of **decline**, which represents the state of the mature tourist destinations. The problems which were sensed in the stagnation stage now manifest themselves clearly, the model of tourism adopted becomes exhausted and it is necessary to redress the situation, to invert the downward trend of the curve. In the face of this situation the mature destinations can opt for various scenarios of growth or decline:
  - Continued decline, due to the passivity of the public and private agents, until there is no longer any solution (Lines D&E in Figure 7.2).
  - Stagnation, due to the application of piecemeal measures which do not attack the root of the problems but only the most evident effects (Line C in Figure 7.2).
  - A radical change of mentality, leading to the adoption of measures which may even entail a new tourism model, based on sustainability and the integration of tourism with the territory, the economy and the local population (Lines A & B in Figure 7.2) (Alvarea and Lourenco, 2005).

This model is compared to the actual development of tourism in the case study area of Sharm in order to predict the city future growth rate.

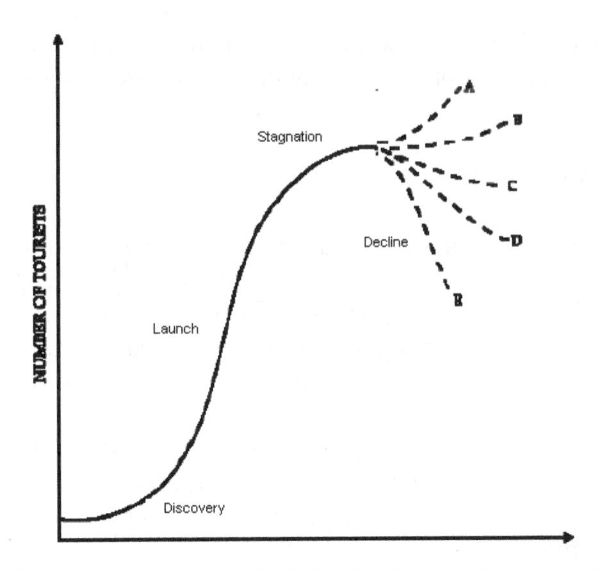

Figure 7.2 Tourism area life cycle model (adapted from (Alvarea and Lourenco, 2005))

Tourism life cycle in the case study area (Sharm)
A lot of development has occurred in the 1980's in Sharm and the city is almost fully developed at the moment where most of the new construction at resorts takes place outside of the city limits (Chapter 2). Figure 7.3 shows actual growth rates in Sharm from 2001 until 2007. The *discovery* period dates back to early 90's but there is no data available for the number of tourists before 2001.

The graph shows continuous growth until the year 2004 (an average growth rate of 20% per annum), a stage of stagnation was observed for the following three years with very slight increases in the number of tourists (6%, 1% and 3% consecutively). This could have been attributed to a terrorist attack in year 2004 and 2005 in Taba (a nearby touristic city) and Sharm consecutively. In 2008, a growth rate of 11% in the number of tourists is expected by the government (XIn, 2006). Figure 7.3 shows that the tourism growth rate in Sharm follows a similar tourism life cycle as explained earlier by Butler.

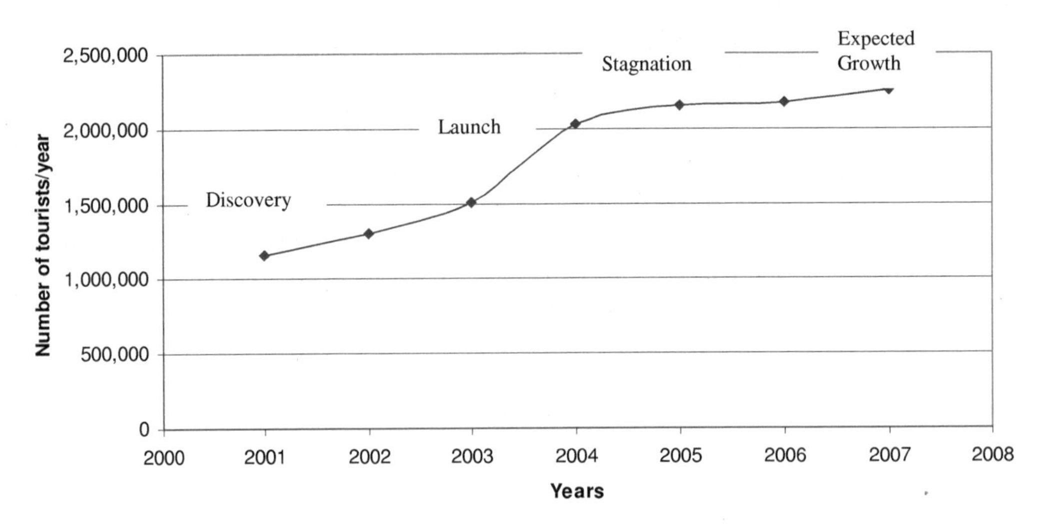

Figure 7.3 Number of tourists in Sharm from 2001 to 2007 (ESIS, 2008; Saad, 2005; WTO, 2007)

From Figure 7.3, the number of tourists in Sharm in year 2007 is 2.25 Million. With 11% expected growth rate in 2008, the number of tourists is estimated at 2.5 Million. This means about 684 additional tourists each day ((2.5-2.25 Million tourists/365 days)). With 65 hotels in Sharm each with 200 rooms in average (Chapter 2), 5 extra rooms are needed at each hotel-assuming equal expansion at each hotel- (684 tourists/65 hotels/2.3 guests per room (Table 7.1)) to accommodate the additional number of tourists (or about 3% growth rate in the number of rooms at each hotel).

The growth rate is multiplied by the parameters in Equation 6.1 to calculate the domestic water demand for years 2 through 5. Using the historical data from the case study area, a 3% growth rate in the contracted-for water supplied was observed each year except for year 3; 0% growth rate was reported.

Contracted-for water supply

Contracted-for water supply (Q_c) is the optimum supply for a hotel ensuring minimum domestic water cost to that hotel (based on average occupancy rate). Q_c is a fraction of the expected maximum water demand. Contracted-for water supply is not the total demand that a hotel requires from a water company. There is also the difference between expected actual demand and contracted-for water supply which results from seasonal variations and spike water demand (due to local holidays, e.g. Christmas, New Year, and Islamic holidays) where occupancy rate is expected to be 100% for most hotels (Chapter 6).

Average daily domestic water demand consists of: 1) total consumption by guests; 2) consumption in swimming pool; 3) consumption of staff during working hours; and 4) consumption of staff housing. Average daily domestic water demand for each month is calculated based on the following equation (Chapter 6):

$$Q_{dom,j} = (SC_g \times f_g \times O_j \times N_r + SC_{p,j} \times A_p + N_s \times (SC_s + SC_{sh})) \times f_{un} \times f_{wdm}$$

Equation 6.1

where $Q_{dom,j}$ is domestic water demand in month j (m^3/d), SC_g is specific consumption per guest ($m^3/cap/d$); f_g is a fraction for number of guests per room; O_j is average occupancy rate in month j (%); N_r is number of rooms in a hotel; SC_p is specific consumption for swimming pool at month j ($m^3/m^2/d$); Ap is total pool area (m^2); SC_s is specific consumption of staff during working hours ($m^3/cap/d$); N_s is number of staff; SC_{sh} is specific consumption of staff housing ($m^3/cap/d$); f_{un} is a factor to estimate unaccounted water in a hotel e.g. cleaning pathways and leakage (%); and f_{wdm} is a factor to account for the level of water demand management inside a hotel (%).Example parameter and model output values are shown in Tables 7.1 & 7.2.

The contracted-for domestic water supply would be calculated as follows:

$$Q_{c,j} = f_O \times Q_{peak,j}$$

Equation 6.3

where $Q_{c,i}$ is contracted-for water supply in month j (m^3/d; or m3/month multiplying by 30 days); $Q_{peak,i}$ is the peak domestic water demand in month j (m^3/d) (calculated using

Equation 6.1 with 100% occupancy rate and highest expected temperature). Q_{peak} does not include the demand for irrigation water; f_O is a fraction of the peak demand (%).

The following equation is used to calculate variations in demand due to the difference between the average occupancy rate and the expected actual occupancy rate in a month:

$$Q_{ed,j} = \max\{(O_{ac,j} - O_j) \times N_r \times N_g \times SC_g, 0\} \quad \text{Equation 7.5}$$

where $Q_{ed,,j}$ is excess water demand due to difference between expected actual and average occupancy rate in month j (m^3/d; or m^3/month multiplying by number of days where occupancy rates exceed average); $O_{ac,j}$ is actual occupancy rate in month j (%); O_j is average occupancy rate in month j (%). $O_{ac,j}$ is assumed as 100% during certain days of the year due to local holidays.

The following equation is used to calculate required water supply in excess of contracted-for water supply:

$$Q_{ec,j} = Abs[\min\{Q_{c,j} - Q_{dom,j} - Q_{ed,j}, 0\}] \quad \text{Equation 7.6}$$

where $Q_{ec,j}$ is required potable water supply in excess of contracted-for supply in month j (m^3/d).

Historical contract data for the case study area are used in this paper. Using Equations 6.1, 6.3, 7.5 and 7.6 (and parameters in Tables 7.1 and 7.2), contracted-for water, excess water demand and required excess potable water supplies are calculated for year 1 and compared to historical data in order to give some guidelines on using the suggested model for future applications. Variation to actual excess water demand is introduced to test impact on optimized RO capacity and expected net benefit.

Table 7.1 Input and built-in parameters used for the demand/supply models calculations

Model Input	Symbol	Unit	Example values	Comments
Fraction for unaccounted-for water	f_{un}	-	0	
Fraction for water demand management level	f_{wdm}	-	0	
Model built-in parameters				Use same values unless other information is available
Factor indicating the percentage of domestic water demand ending up as sewage	f_{dw}	%	80	(Khaled, 2008; Mohsen, 2007)
Fraction for number of guests per room	f_g	-	2.3	Average from nine surveyed hotels in Sharm (Chapter 5)
Optimized fraction of peak demand	f_o	%	100	Value of f_o before iteration
Fraction for number of staff per room	fs	-	1.05	(Chapter 5)
Specific water consumption of guests	SC_g	m^3/cap/d	0.4	Average from nine surveyed hotels in Sharm (Chapter 5)
Specific water consumption of irrigated land	SC_{gr}	m^3/m^2/d	0.01	From an empirical equation of average monthly temperature[a]
Specific water consumption of swimming pool – if using fresh, not saline water	SC_p	m^3/m^2/d	0.01	From an empirical equation of average monthly temperature (Chapter 5)
Specific water consumption of Staff housing	SC_{sh}	m^3/cap/d	0.25	Average from nine surveyed hotels in Sharm (Chapter 5)
Specific water consumption of staff	SC_s	m^3/cap/d	0.03	Average from nine surveyed hotels in Sharm (Chapter 5)
Average yearly temperature	T	°C	29	Average monthly temperature varies from 18 in January to 42°C in August

[a]Equation for calculation of green irrigated area specific consumption SC_{gr}: 0.0047+0.00112T; where SC_{gr} is in m^3/m^2/d and T is temperature in °C (Data collected from 9 hotels in Sharm El Sheikh)

Table 7.2 Hotel characteristics and the demand/supply models output values in year 1

	Symbol	Unit	Hotel 1	Hotel 2	Hotel 3	Hotel 4	Hotel 5	Comments
No. of rooms	N_r	-	107	458	99	86	500	
Area of pool	A_p	m^2	800	800	800	800	800	
Average monthly occupancy rate	O	%	57%	73%	57%	75%	56%	
Staff housing within premises			yes	yes	yes	yes	yes	Hotels can have staff housing within premises or outside
Staff number	N_s	-	112	480	104	90	525	Ratio of staff to rooms f_s (Table 7.1) multiplied by no. of rooms in a hotel
Domestic water demand	Q_{dom}	m^3/d (m^3/month)	90 (2,700)	447 (13,410)	93 (2,790)	87 (2,610)	406 (12,180)	In the month of January as an example calculated using Eq. 6.1
Contracted-for water supply	Q_c	m^3/d (m^3/month)	86 (2,581)	373 (11,189)	91 (2,727)	88 (2,652)	403 (12,099)	Using Eq. 6.3 Total equals 1,041 m^3/d (31,250 m^3/month or 375,036 m^3/yr)-matches historical data for year 1
Fraction of peak demand	f_o	%	60	65	58	75	65	
Domestic peak demand	Q_{peak}	m^3/d (m^3/month)	143 (4,290)	574 (17,220)	157 (4,710)	118 (3,540)	620 (18,600)	Calculated using Eq. 6.1 using temperature of 42°C (August temp) and 100% occupancy rate
Excess water demand	Q_{ed}	m^3/month	210	570	195	100	1,015	In the month of January as an example, using Eq. 7.5 and parameters in Table 7.1, e.g. for hotel 1 (100%-57%)*107*2.3*0.04*5 days in January during Christmas and New Year holidays
Excess water supply	Q_{ec}	m^3/month	343	2,782	249	71	1,110	Total equals 84,684 compared to actual extra water supply reported by company of 84,619 m^3/yr. Using Eq. 7.6: abs (2581-2700-210)

Economic model

The economic model calculates the capital and production costs of RO desalination as well as expected revenue (based on prices during high and low demand periods).

Cost Model
Most economists in theory and practice assume considerable economies of scale in water production and supply. Nevertheless, no consensus seems to exist with respect to the range of the production function over which economies of scale do hold (Sauer, 2005). Unit production cost consists of amortized capital cost in addition to operation and maintenance (O&M) costs including labour and administration, chemicals, energy and maintenance including membrane replacement.

Unit costs are a function of utilization. Utilization is defined as actual production of the RO desalination plant divided by the RO plant capacity. 19 RO plants in Sharm were surveyed (owned and/or operated by Ridgewood Egypt) and data were collected including unit production costs, O&M costs, energy consumption rates, contracted-for water supply, and utilization. The RO desalination plants ranged in size from 450 to 3,500 m^3/d. Figure 7.4 shows variation of unit costs as a function of utilization over the year for one of the surveyed plants.

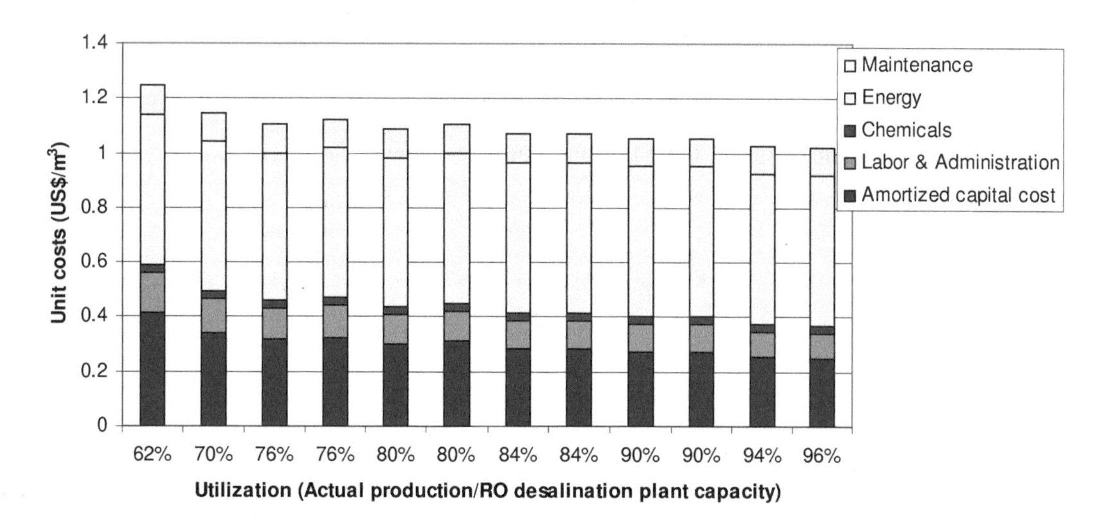

Figure 7.4 RO actual unit costs as a function of utilization for the example RO plant in Sharm

Amortized capital costs are inversely proportional to utilization. Chemicals and maintenance unit costs are not related to utilization. Labour and administration costs decreases as utilization increases. This indicates that cost of labour and administration is paid by the water company independent of production. Brine is disposed into wells. The cost of brine disposal is included in the capital cost (well construction and pipeline).

Hafez and El Manharawy (2002) state that the capital costs for plants size ranging from 250 m^3/d to 50,000 m^3/d show economies of scale, however, O&M costs are proportional to the plant size. In this paper, unit capital cost is introduced using an empirical equation to illustrate economy of scale with increased capacity:

$$C_c = 603\exp^{-2\times10^{-5}Q_w} \qquad \text{Equation 7.7}$$

Where C_c is unit capital cost (US\$/m^3) and Q_w is RO desalination plant capacity (m^3/d). Unit capital cost is multiplied by the RO plant capacity to get total capital costs.

Capital costs are amortized over the lifetime of the project using annuity factor *a* determined from the following equation (Chapter 3):

$$a = \frac{i\times(1+i)^n}{(1+i)^n - 1} \qquad \text{Equation 3.1}$$

Where *i* is the interest rate (8 %) and *n* is the economic plant life (10 years lifetime). To convert annual capital cost to a unit cost, it can be further divided by the annual designed plant output (RO plant capacity multiplied by 365 days).

O&M costs from the surveyed 19 RO desalination plant show variation of desalination O&M costs with plant size (Figure 7.5). The O&M costs from the surveyed RO plants showed clear economies of scale with increased plant size contrary to the findings by Hafez and Manharawy (2002). A multiple linear regression (MLR) analysis was carried out using data from the 19 surveyed RO plants to obtain unit cost equations for the different components of O&M as a function of utilization and capacity to be used for the cost model. For the MLR, the following equation was used:

$$c = b + d\times U + e\times Q_w \qquad \text{Equation 7.8}$$

where *c* is unit cost of different O&M components (US\$/m^3); *U* is utilization factor (%); Q_w is RO desalination plant capacity (m^3/d); *b, d,* and *e* are constants.

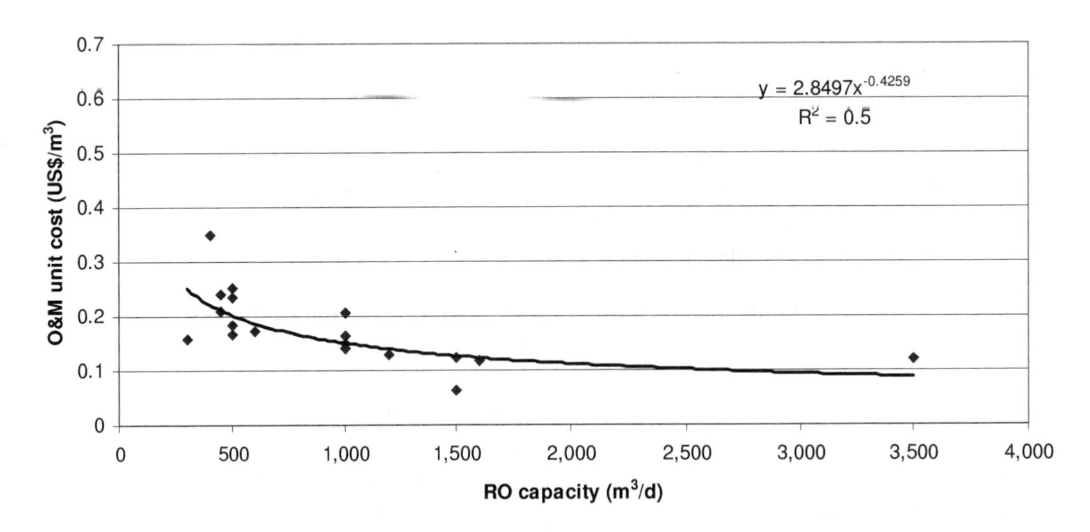

Figure 7.5 Actual O&M costs for the 19 RO plants in Sinai (Girgis, 2008)

The MLR analysis yielded the following values for the constants *b, d,* and *e* for labour and administration, chemicals, and maintenance respectively.

Labour and administration:

$$c_l = 0.18 - 0.1U - 1.2\times10^{-5} Q_w \qquad \text{Equation 7.9}$$

Constant *b* (0.18) represents unit labour cost that is more or less fixed and independent of utilization or capacity of RO desalination plant. Including utilization improved the correlation with 14% while capacity yielded a correlation of 46% yielding a total correlation of 60%.

Chemicals:

$$c_{ch} = 0.02 + 0.004U - 3.6\times10^{-6} Q_w \qquad \text{Equation 7.10}$$

Constant *b* (0.02) represents unit chemicals cost that is more or less fixed and independent of utilization or capacity of RO desalination plant. Including utilization improved the correlation by only 0.6% while capacity yielded a correlation of 24%. Chemical cost was taken as a fixed value (average chemical cost of the 19 surveyed plants).

Maintenance:

$$c_m = 0.12 - 0.05U - 1.15\times10^{-5} Q_w \qquad \text{Equation 7.11}$$

Constant *b* (0.12) represents unit maintenance cost that is more or less fixed and independent of utilization or capacity of RO desalination plant. Including utilization improved the correlation with 0.2% while capacity yielded a correlation of 47%.

Unit cost of energy is a function of cost of energy and energy consumption rate so it was taken as a fixed value (Table 7.3).

The cost of water supplied is calculated for each year over the lifetime of the project. In order to compare cost and revenue over the lifetime of the project and not just for each year separately, the yearly values are converted into a single equivalent value: present value (i.e. what it is worth today to pay or gain a certain amount of money after a number of years). Yearly costs/revenue is changed to present value using the following equation:

$$V_o = \frac{V_y}{(1+i)^y} \qquad \text{Equation 7.12}$$

where V_o is the present worth of cost or revenue (US$/yr); V_y is the sum of monthly cost or revenue in year *y* (US$/yr) (Cost is the sum of unit amortized capital cost using Equations 7.7 and 3.1 multiplied by monthly designed RO plant output; and unit O&M costs multiplied by actual monthly production/supply (Q_s) using Equations 7.8 through 7.11 in addition to energy costs (Table 7.3); revenue is calculated using Equation 7.13); *i* is the interest rate (8 %); *y* is the number of years (1 to 5 years).

Pricing structure for water from RO plants in Sharm

Effective water pricing and provision of regulatory and economic incentives to reduce demand are important for sustainable water resources management. Real success and viability for water resources management plan depend on proper enforcement of charges and monitoring of users (Seppala and Katko, 2003). The current trend for the three private water companies in the case study area is to charge the same price irrespective of the amount of water consumed to users having contracts with the company. A higher price is only applied to users with no contracts with the water company.

This trend provides no incentive to customers (with contracts) to reduce their consumption or abide by the contracted-for amount (Chapter 6). In this paper, a pricing structure is proposed which is a function of water consumption. The proposed pricing structure ties consumption to prices, i.e. the more the user consumes water, the higher the water price. This scheme can assist in reducing demand in excess of contracted-for water supply aiding the water company to fulfil its contractual commitments to all users.

Most empirical estimates of residential water demand report inelasticity, i.e. a 1% increase in price results in less than 1% decrease in consumption (Xu, 2002). The suggested pricing structure has to be tested in practice to observe associated impact on demand, if any. The proposed pricing structure used for the cost model and the expected revenue from water sales for the water company is calculated as follows:

$$R_y = \sum_{j=1}^{12} (Q_{c,j} \times P_p + \max((Q_{s,j} - Q_{c,j}),0) \times P_{EP})$$

If $Q_{c,j} < Q_{s,j} \leq 1.1Q_{c,j}$ then $P_{EP} = vP_p$ else
If $1.1Q_{c,j} < Q_{s,j} \leq 1.3Q_{c,j}$ then $P_{EP} = wP_p$ else
If $Q_{s,j} > 1.3Q_{c,j}$ then $P_{EP} = z\,P_p$ Equation 7.13

where $Q_{c,j}$ is contracted-for water supply in month j (m^3/d); P_p is price of contracted-for water supply (US$/$m^3$); $Q_{s,j}$ is water supply in month j (m^3/d) (summation of $Q_{c,j}$ and all or part of $Q_{ec,j}$ depending on the optimization solution); P_{EP} is price of water supply in excess of contracted amount (US$/$m^3$); v,w,z are constants (with values equal to or larger than 1, and with $v<w<z$). Values of v,w,z are 1.16, 1.2 and 1.3 respectively. Example parameter values are shown in Table 7.3.

Table 7.3 Input parameter values for the economic model

Model Input	Symbol	Unit	Example values	Comments
Water selling price for contracted-for water supply and excess water supply, respectively	P_p, P_{EP}	US$/$m^3$	1.07, 1.25	Average Selling price was fixed for this plant over the period from 2004 to 2008
Energy cost	c_e	US$/$m^3$	0.05	Average cost from 19 surveyed plants
Energy consumption rate	-	kWh/m^3	7	Average consumption from 19 surveyed plants
Chemicals cost	c_{ch}	US$/$m^3$	0.02	Average cost from 19 surveyed plants

For the optimization model, the actual selling prices (P_p and $P_{EP,}$ for users with contracts and users without contracts respectively) from the example RO plant are used, the actual net benefit is then compared to the expected net benefit if the suggested pricing structure is adopted (Table 7.4). For this RO plant, most of the water sold in excess of contracted-for amount is for outside sales (i.e. users without contracts).

Application of the dynamic optimization model on the case study area

An example RO plant operated by Ridgewood Egypt was used for the optimization modeling in this paper. Using the cost functions for the capital and O&M costs, the cost model calculated similar values to the actual net benefit for the example RO plant. Figure 7.6 shows comparison between the actual water production costs from the example RO plant in Sharm and water production costs calculated using the cost model.

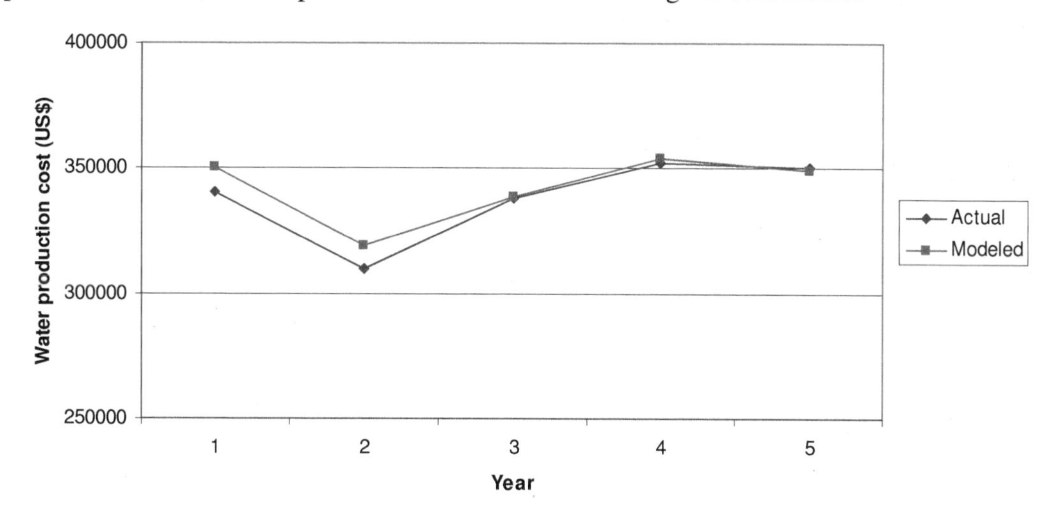

Figure 7.6 Comparison between actual and modeled water production costs for the example RO plant in Sharm

Capacity expansion was estimated for 5 years (assumed planning horizon of the project) and compared to actual figures. Table 7.4 and Figure 7.7 summarize the actual contract and excess water supply, actual plant capacity and results of the optimization of the plant capacity expansion using the current pricing structure on one hand and the suggested pricing structure in this paper on the other (using historical data) over the 5-year period. The following numerical example explains the usage of Equations 7.1 through 7.4 to solve for the optimization problem:

Max benefit at end of year 5 which is equivalent to $F_5(s_6)$ = US$ 737,326 , where s_6 was equal to 1,670 m^3/d. Backtracking to obtain best path:

Total net benefit at end of year 4 $f_4(s_5)$ = 737,326-165,381 (net benefit from yr 5 only) = US$ 564,225; where s_5 was equal to 1,650 m^3/d.

Total net benefit at end of year 3 $f_3(s_4)$ = 56,4225-168,457 (net benefit from yr 4 only) = US$ 403,178; where s_4 was equal to 1,520 m^3/d.

Total net benefit at end of year 2 $f_2(s_3)$ = 403,178-135,769 (net benefit from yr 3 only) = US$ 266,455; where s_3 was equal to 1,320 m^3/d.

Total net benefit at end of year 1 $f_1(s_2)$ = 266,455-119,181 (net benefit from yr 2 only) = US$ 148,228; where s_2 was equal to 1,290 m^3/d.

Table 7.4 Capacity optimization for the example RO plant in Sharm

Year[a]	Contracts value Q_c (m^3/yr)	Excess supply Q_{ec} (m^3/yr)	Actual capacity Q_w (m^3/d)	Optimized capacity[b] Q_w (m^3/d)	Optimized capacity[c] Q_w (m^3/d)	Actual benefit B (US$)	Optimized benefit[b] B (US$)	Optimized benefit[c] B (US$)
1	375,036	84,619	1,450	1,240	1,290	137,054	144,013	148,228
2	385,260	43,427	1,425	1,270	1,320	108,616	120,152	119,181
3	384,276	92,878	1,550	1,520	1,520	123,797	128,398	135,769
4	397,524	146,035	1,647	1,650	1,650	160,124	154,895	168,457
5	409,449	150,416	1,730	1,690	1,670	141,990	153,411	165,689
Total						**671,582**	**700,928**	**737,324**

[a] The net benefit numbers calculated for each year is converted to the present value using Eq. 7.12

[b] Using pricing structure originally used by company (contracted-for water selling price is P_p and excess water is P_{EP}, excess demand is mainly from users without contracts)

[c] Using modified pricing structure (revenue from contracted-for water sales (at price P_p) and excess water sales is calculated (at price P_{EP}) using Eq. 7.13, assuming excess demand is from users with contracts only)

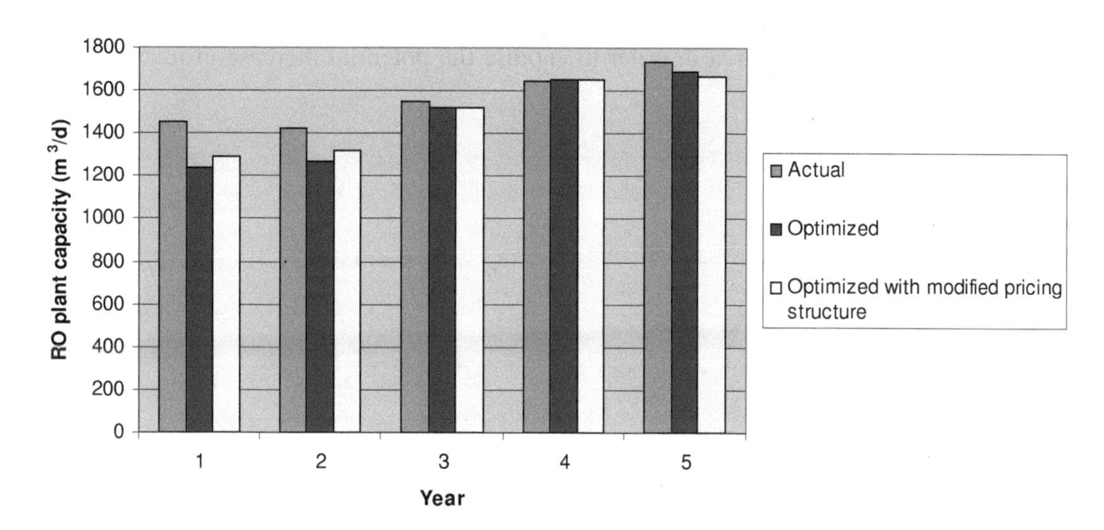

Figure 7.7 Comparison between actual plant capacity for the example RO desalination plant and results obtained from the optimization problem using current and modified pricing structures

The optimized RO capacity does not necessarily satisfy the total required water supply (equal to contracted-for water and excess water supplies Q_c+Q_{ec}) (Figure 7.8). Using the current pricing structure as the water company, for years 1 and 2, the maximum net benefit could have been obtained by decreasing the installed capacity from 1,450 and 1,425 m^3/d to 1,240 and 1,270 m^3/d respectively. For years 3, 4, and 5 the actual installed capacities were only slightly different from the values calculated by the model.

Using the modified pricing structure, for years 1 and 2, the optimum capacity is higher than the one calculated with the current pricing structure, i.e. the increased revenue from sales at a higher price offset the increased cost of installing and operating a higher capacity plant. For years 3 and 4, the optimum capacity is the same in both cases (using current and modified pricing structures) while for year 5, the optimum capacity using the modified pricing structure is slightly lower than using the current pricing structure. Overall, by adopting the suggested pricing structure in the paper, the maximum total net benefit is higher than what could have been achieved using the current pricing rates.

Due to the price inelasticity explained earlier, it is assumed that the pricing structure will not cause the reduction of water demand. It should be noted that the demand function is, to a large extent, defined by the behaviour of hotel guests. However, these guests are not aware of the water prices, nor do they pay according to their water consumption. A pricing structure can only have an effect on demand as far as demand is influenced by the decision makers (who are aware of the prices), i.e. the hotel managers and some of the staff. Therefore, the water company operates in a market with a limited number of relatively large customers, and in such a market, the concept of elasticity is less applicable; real change should be negotiated to reduce demand such as shifting to saline water for swimming pools, and less-water consuming devices.

Varying the excess water demand from actual (historic) values, over the 5-year period, yielded a 2% increase in total net benefit for a 5% yearly increase in excess water demand (for instance due to an increased influx of tourists). Using the optimization model, this increase necessitated additional RO plant capacity in years 1 & 5 but not in years 2, 3, and 4. A comprehensive updated marketing analysis is essential for the water company to keep track with any governmental promotional efforts to attract tourism, or any planned expansion by hotels in the area in order to capture the potential increase in market share and maximize benefits.

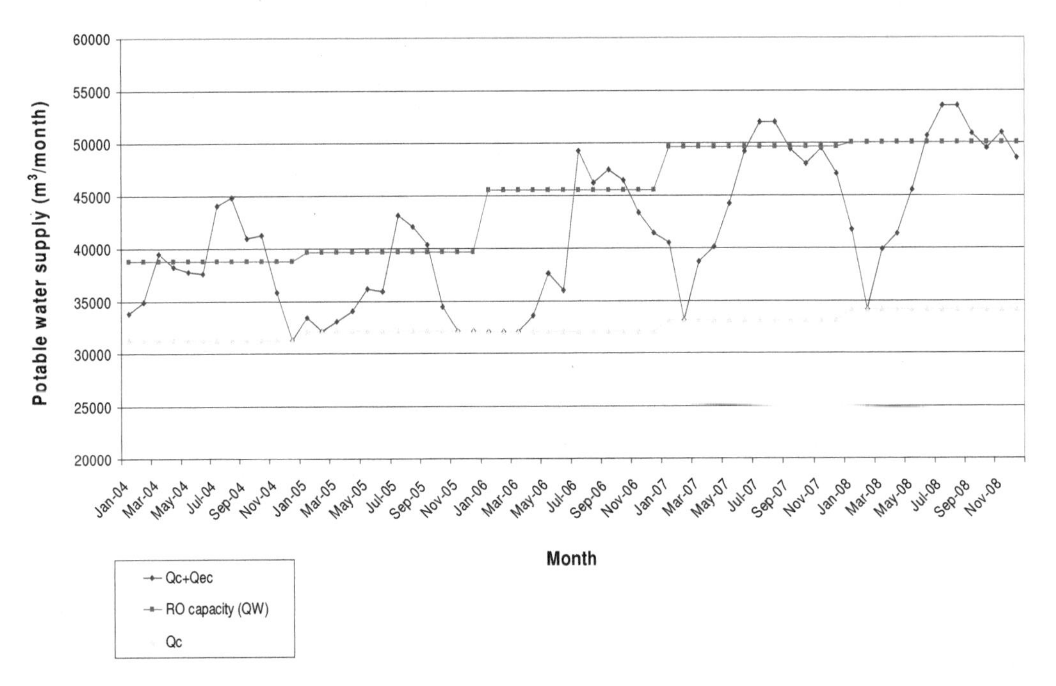

Figure 7.8 RO desalination plant capacity, contracted-for water supply and excess water supply for an example RO plant in Sharm from 2004 to 2008

Conclusions and recommendations

A model was developed using Excel macros to perform dynamic programming to solve an optimization problem with the objective function to maximize the present value of total net benefits over the lifetime of the RO desalination plant. The aim of the optimization is to solve for capacity expansion. The model can be used to test different scenarios or changes in parameters to capture time-variant tourism demand and price uncertainties on investment decisions.

This study focuses on tourism dominated arid coastal regions, using Sharm El Sheikh (Sharm) in South Sinai, Egypt, as an example. 19 RO plants in Sharm were surveyed and data were collected including unit production costs, O&M costs, energy consumption rates, contracted-for water supply, and utilization.

Using the collected data, cost functions were developed for O&M costs as a function of utilization and plant capacity. The cost model calculated similar values to the actual net benefit for one of the surveyed RO plant taken as an example. Unit production costs of RO desalination plants vary according to the degree of operation of the plant. When the plant is 100% operational, the unit production cost includes amortized cost on capital investment, in addition to costs of energy, labour, chemicals, and maintenance including membrane replacement. Even when plant is partially operational, there are costs which are still incurred: amortized cost on capital, and labour and administration. This fact has to be taken into consideration when calculating costs of RO desalination and when deciding on plant capacity and whether peak demand should be satisfied in order to maximize net benefit.

With an RO plant designed to satisfy only the contracted-for water supply, the water company would be missing out on potential benefits that could have been obtained selling water in periods of high demand. On the other hand, sizing the RO plant to produce water to cover for peak demand means incurring additional costs as well as having the plant partially idle during periods of average or low demand. Using the optimization model, the maximum net benefit is obtained with a smaller installed capacity than the actual case.

A modified pricing structure is suggested in the paper that ties the water selling price to consumption in an effort to reduce demand in excess of contracted-for water supply aiding the water company to fulfil its contractual commitments to all users. Pricing excess water over contracted-for supply at higher rates than usual may retain an incentive to promote water conservation (Meij *et al.*, 2005) as it allows the water company to efficiently control consumption and satisfy their agreements with contracted-users. However, price elasticity has to be taken into consideration to determine the impact of price change on water demand. It should be noted that the demand function is, to a large extent, defined by the behaviour of hotel guests. However, these guests are not aware of the water prices, nor do they pay according to their water consumption. A pricing structure can only have an effect on demand as far as demand is influenced by the decision makers (who are aware of the prices), i.e. the hotel managers and may be some of the staff. Real change should be negotiated to reduce demand such as shifting to saline water for swimming pools, and less-water consuming devices.

It should be noted that planning for the future always involves uncertainties due to variability of the situation, e.g. changes in demand and consumption patterns and economic conditions. A comprehensive updated marketing analysis is essential for the water company to keep track with any governmental promotional efforts to attract tourism, or any planned expansion by hotels in the area in order to capture the potential increase in market share and maximize benefits.

Acknowledgements
We would like to thank Mr. Zaki Girgis, CEO of Ridgewood Egypt for critically reviewing this paper and for kindly providing valuable data.

8 Integrating wastewater reuse in water resources management for hotels in arid coastal regions - Case study of Sharm El Sheikh, Egypt

Abstract

Hotels in arid coastal areas use mainly desalinated water (using reverse osmosis) for their domestic water supply, and treated wastewater for irrigating green areas. Private water companies supply these hotels with their potable and non-potable water needs. There is normally a contractual agreement stating a minimum amount of water that has to be supplied by the water company and that the hotel management has to pay for regardless of its actual consumption ("contracted-for water supply").

This paper describes a model to optimize the contracted-for irrigation water supply with the objective function to minimize the total water cost to hotels. It analyzes what the contracted-for irrigation water supply of a given hotel should be, based on the size of the green irrigated area on one hand and the unit prices of the different types of water on the other hand. An example from an arid coastal tourism-dominated city is presented: Sharm El Sheikh (Sharm), Egypt.

This paper presents costs of wastewater treatment using waste stabilization ponds which is the prevailing treatment mechanism in the case study area for centralized plants, as well as aerobic/anaerobic treatment used for decentralized wastewater treatment plants in the case study area.

Contracting for full peak irrigation demand is the highest total cost option. Contracting for a portion of the peak irrigation demand and complementing the rest from desalination water is a cheaper option. A better option still is to complement the excess irrigation demand from the company that treats and sells wastewater, if available, or from another wastewater treatment company at a higher cost (but at a cost cheaper than that of desalination water) mainly due to the high demand season and the additional cost of trucking. In some cases however, like in Sharm, the amount of treated wastewater is limited and variable during the year and some hotels have no choice but to partially use desalinated water for irrigation.

A conscious strategy for water management should rely solely on treated wastewater on-site. This can be achieved by: increasing the efficiency of the irrigation system, reducing the area of high-water consuming plantation (e.g. turf grass) and/or shifting to drought resistant plants including less water-consuming or salt tolerant turf grass.

Keywords: Tourism; water demand and supply; treatment; irrigation; water contracts.

This chapter is based on:
Lamei, A., van der Zaag, P. and Imam, E. (submitted) Integrating wastewater reuse in water resources management for hotels in arid coastal regions Case study of Sharm El Sheikh, Egypt. *Water Sci. & Tech.*

Introduction

Treated wastewater can advantageously find many applications including agricultural, industrial, urban reuse and groundwater recharge. However, wastewater reuse requires effective treatment and measures to protect public health and the environment at a feasible cost (Sipala *et al.*, 2003). Wastewater re-use increases water supply, saves the environment from the damaging effect of dumping wastewater and thus not only justifies the allocated resources but also supports sustainable development (Khouzam, 2003). Wastewater reuse can result in the maintenance of higher quality water.

In arid coastal regions, seawater or brackish groundwater are often the only water sources for domestic purposes (treated by reverse osmosis desalination) while treated wastewater is used for irrigation. In case there is a shortage of treated wastewater during certain periods of the year depending on the decreased use of potable water, desalinated water is used to complement the supply for irrigation purposes. In many of these coastal arid regions, the main economic activity is tourism, implying a concentration of wastewater generation from hotel operations (Gossling, 2000). In these situations, both desalination and wastewater treatment are mainly managed by the private sector, whether on-site for individual hotels or centralized for a group of hotels.

The supply of potable and non-potable water from centralized plants is done under certain contractual arrangements with the tourism sector. The most commonly used contract type is the BOO (Build, own, and operate) contract (Hafez and El Manharawy, 2002). The water company (contractor) will construct and operate the plant, with contracts for potable or non-potable water specifying a minimum agreed-upon water quantity (Chapter 6). The hotel management will have to pay for that water (potable and non-potable) supply whether it is actually consumed or not.

Water demand in excess of the contracted amount is supplied by the water company only if it is available. Since the water company usually has contracts for nearly the total capacity of its water production plant, no excess water is available in most cases, especially in hot months when irrigation demands are intense. If the hotel's actual water demand is higher than the contracted-for water then it may have to buy water from other sources normally at higher prices (due to increased demand and trucking cost). If treated wastewater is insufficient, desalinated water (if available) may have to be purchased (at a higher cost) to satisfy irrigation requirements.

The focus of this paper is on treated wastewater contracts for hotels. An earlier research discussed the optimum contracted-for potable water supply (Chapter 6). Similar to domestic water demand, irrigation water demand fluctuates during the year depending on climate. Hotel managers usually calculate the contracted-for supply as a percentage of their peak daily irrigation water demand. In Sharm El Sheikh (Sharm), for example, the peak daily irrigation water demand is expected to occur in August where the ambient temperature is at its maximum (approximately 40 °C) and rainfall is negligible (Chapter 2).

Hotels have the choice of contracting for the full peak daily irrigation demand (i.e. 100%, though it is not fully used throughout the year), or for a lesser amount. If the former, they either do not use the full contracted amount (though they pay for it) or they over irrigate their landscape with the contracted-for water. On the other hand, if they contract for less than the full peak demand, they complement the unsatisfied demand during the hottest months with expensive desalinated water. As hotel managers do not usually account for irrigation water demand when considering contracts for domestic water supply, the bought desalinated water will be obtained at a higher price than the regular contracted-for desalinated water price. A particular case of interest is the case of hotels with golf courses.

This paper discusses a model which optimizes the contracted-for irrigation water supply with an objective function to minimize the overall total cost of domestic and irrigation water considering different scenarios for the supply of irrigation water.

The paper focuses on arid coastal regions, using Sharm in Egypt as an example. Similar contractual arrangements are practiced by water companies supplying potable water in coastal regions in Egypt (Khalil, 2004) and also worldwide in areas where the tourism industry is the dominant water user, e.g. in Greece and Spain (Avlonitis *et al.*, 2002; Gasco, 2004; Khalil, 2004).

According to the authors' knowledge, there is no reference in the literature to optimization of contractual supply for treated wastewater. However, Zein (2006) has studied the wastewater management system in Egypt from the institutional point of view but the study did not present data on treatment costs. Sipala *et al.* (2003) calculated unit production costs for wastewater treatment for different technologies for reuse. The authors mention that wastewater treatment costs are not well documented. Their model did not include land-based treatment systems for wastewater which is the major wastewater treatment method in the case study area.

This paper presents the optimization model as well as data from the case study area including wastewater inflows, the costs of wastewater treatment using waste stabilization ponds which is the prevailing treatment mechanism for centralized plants, and the costs of aerobic/anaerobic treatment used for the decentralized wastewater treatment plants (Chapter 2).

Contracted-for irrigation water supply optimization model

Optimization model

The objective function of the optimization model is to minimize the yearly cost of domestic and irrigation water for the hotel while satisfying the water needs. The equation can be expressed as follows:

Minimize

$$C_{T,y} = \sum_{j=1}^{12} P_P \times Q_{C,j} + P_{EP} \times \max(Q_{dom,j} - Q_{C,j}, 0) + P_{TWW} \times f_{dw} \times Q_{dom,j} + P_{WW} \times Q_{IC,j}$$

$$+ \min(P_{EWW}, P_P, P_{EP}) \times Q_{IEC,j}$$

Decision variable: f_o Equation 8.1

where $C_{T,y}$ is the total cost for domestic and irrigation water supply in year y(US\$/yr); P_P is the selling price of potable water (US\$/m^3); $Q_{c,j}$ (see Equation 6.3) is contracted-for potable water supply in month j (m^3/month; daily values multiplied by 30 days); P_{EP} is the selling price of potable water in excess of contracted-for water supply (US\$/m^3); $Q_{dom,j}$ (see Equation 6.1) is domestic water demand in month j (m^3/month; daily values multiplied by 30 days); P_{TWW} is the unit cost of treating wastewater within the premises of the hotel (US\$/m^3); f_{dw} is a factor indicating the percentage of domestic water demand ending up as sewage; P_{WW} is the selling price of treated wastewater (US\$/m^3); $Q_{IC,j}$ is contracted-for irrigation water supply in month j (m^3/month; daily values multiplied by 30 days) (see Equation 8.2); P_{EWW} is the selling price of treated wastewater in excess of contracted-for water supply (US$/$m^3$); $Q_{IEC,j}$ is irrigation water demand in excess of contracted-for irrigation water supply in month j (m^3/month; daily values multiplied by 30 days); f_O is a fraction of the peak demand.

Example values for the different selling prices are provided in Table 8.1. In order to solve the optimization model, both the domestic and irrigation water demand have to be calculated.

Domestic water demand
The fresh water demand of hotel is a function of the number of hotel rooms, probable occupancy rate, number of staff working and/or living in a hotel, presence and size of swimming pool, kitchen, laundry, etc. The following equation calculates the daily domestic water demand for hotels (Chapter 6):

$$Q_{dom,j} = (SC_g \times f_g \times O_j \times N_r + SC_{p,j} \times A_p + N_s \times (SC_s + SC_{sh})) \times f_{un} \times f_{wdm}$$

Equation 6.1

where $Q_{dom,j}$ is domestic water demand in month j (m^3/d), SC_g is specific consumption per guest (m^3/cap/d); f_g is a fraction for number of guests per room; O_j is average occupancy rate in month j (%); N_r is number of rooms in a hotel; SC_p is specific consumption for swimming pool at month j (m^3/m^2/d); Ap is total pool area (m^2); SC_s is specific consumption of staff during working hours (m^3/cap/d); N_s is number of staff; SC_{sh} is specific consumption of staff housing (m^3/cap/d); f_{un} is a factor to estimate unaccounted water in a hotel e.g. cleaning pathways and leakage (%); and f_{wdm} is a factor to account for the level of water demand management inside a hotel (%).
Example values for these parameters are provided in Table 8.2.

The contracted-for potable water supply would be calculated as follows:

$$Q_{c,j} = f_O \times Q_{peak,j}$$ Equation 6.3

where $Q_{c,j}$ is contracted-for water supply in month j (m^3/d); $Q_{peak,j}$ is the peak domestic water demand in month j (m^3/d) (calculated using Equation 6.1 with 100% occupancy rate and highest expected temperature, in the case of Sharm during the month of August); f_O is a fraction of the peak demand.

Irrigation water demand
The irrigation water demand is a function of the irrigated green area. The green area could vary from a golf course, large green area or small green area. In Egypt, golf course irrigation is one of the fastest growing reuse applications due to its high water consumption (Lazarova *et al.*, 2001). Specific irrigation water consumption was calculated using CropWat 4. CropWat 4 is a model developed by FAO (Food and agriculture organization) to calculate water irrigation requirements depending on local climate for different crops including grass.

The software is used in corporation with a database Climwat which includes data from 144 countries: max and min temperature, mean daily relative humidity, sunshine hours, wind speed and precipitation.

The contracted-for irrigation supply would be calculated as follows:

$$Q_{IC,j} = f_O \times Q_{Ipeak,j}$$ Equation 8.2

where $Q_{IC,j}$ is contracted-for irrigation water supply in month j (m^3/d); $Q_{Ipeak,j}$ is the peak irrigation water demand in month j (m^3/d) (calculated using CropWat 4 with highest expected temperature, in the case of Sharm during the month of August); f_O is a fraction of the peak demand.

Case study area

Wastewater treatment and reuse in case study area

In Sharm, there are currently no major private wastewater treatment companies selling treated wastewater back to hotels except for one company (Al Montaza) established in late 1995 which has a network connected to a group of 15 hotels. The company collects the hotels' wastewater for free and sells it back after treatment to the same hotels at a fee.

Most other hotels have on-site wastewater treatment plants; for some small hotels, wastewater is discharged to a government-owned wastewater treatment plant where it is treated (primary treatment) and piped to irrigate 40 hectares of tree plantations owned by the local municipality. Figure 8.1 shows a graph for wastewater inflow to the government-owned waste water treatment plant (capacity 15,000 m^3/d) in Sharm. The maximum inflow occurs in August and September.

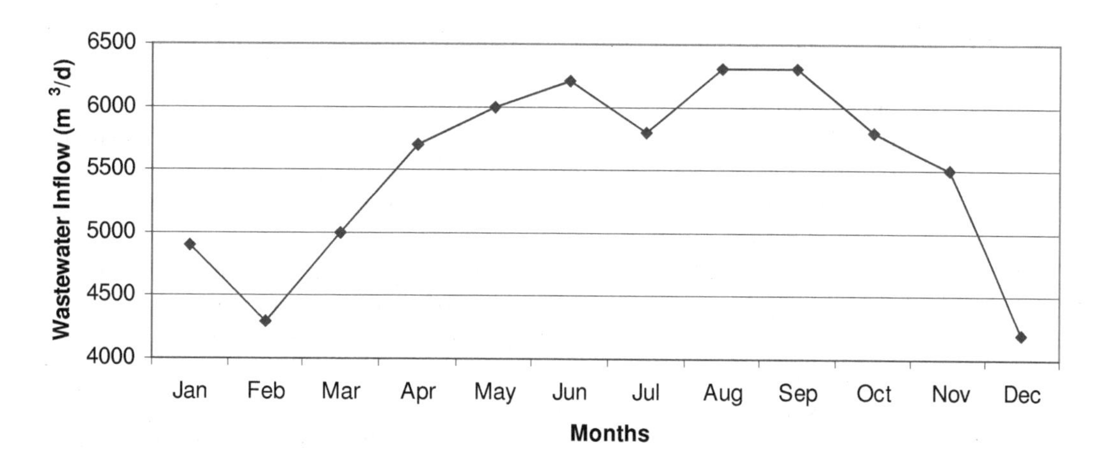

Figure 8.1 Wastewater inflow to the government-owned wastewater treatment plant in Sharm

The government-owned wastewater treatment plant is a waste stabilization pond which is composed of two trains. During maintenance, one train can be disconnected and treatment takes place in the other one only. Before going to the collection pond, chlorine is injected. Treated effluent goes then to distribution: by a pipeline to the forest plantation, and by trucks for landscape irrigation downtown (public landscape).

Figure 8.2 shows unit capital costs of five government-owned wastewater treatment plants in Sinai (all stabilization ponds). For the government-owned wastewater treatment plant in Sharm, the average O&M cost is about 0.2 US$/$m^3$ including maintenance and spare parts, labour and chemicals (chlorine). The unit production cost of treated wastewater is calculated to be about 0.25 US$/m^3 based on 30 years lifetime and 8% interest rate (Chapter 2).

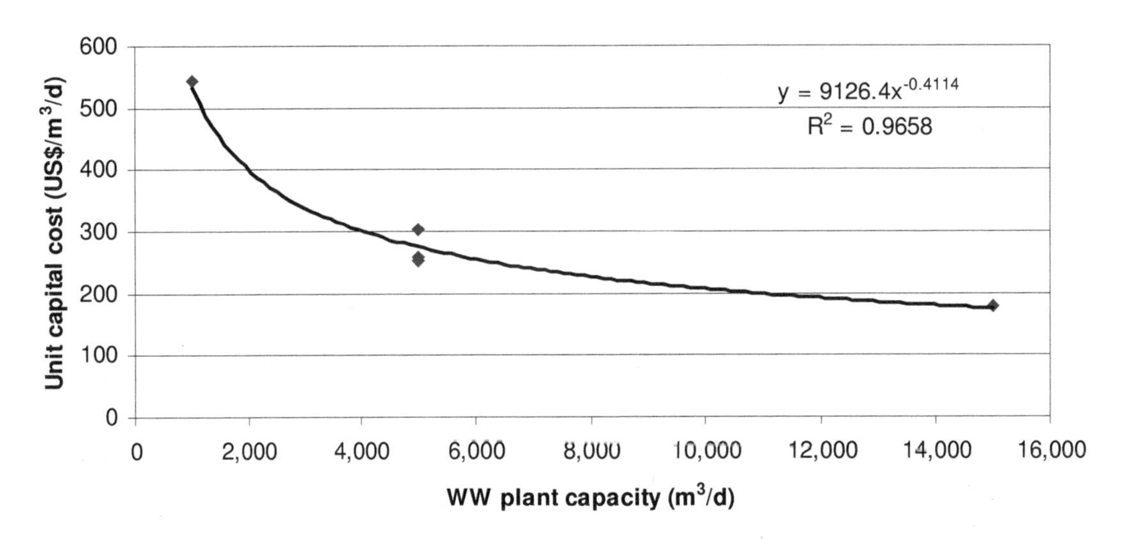

Figure 8.2 Unit capital costs of five government-owned wastewater treatment plants in Sinai (all stabilization ponds)

Example values for the optimization model from Sharm

Six scenarios were considered in the optimization of irrigation contracted-for water supply:

- Scenario 1: All irrigation water demand is covered by the contract (i.e. contracting for 100% of the expected peak irrigation water demand); no wastewater treatment on-site.
- Scenario 2: Excess irrigation water is covered by potable water from desalination at high price (i.e. excess irrigation demand was not accounted for when the potable water demand contract was prepared, therefore, any extra needed water has to be purchased at P_{EP}); no wastewater treatment on-site.
- Scenario 3: Excess irrigation water is covered by potable water from desalination at the contracted price P_p (i.e. excess irrigation demand was taken into consideration when the contract for potable water was prepared); no wastewater treatment on-site.
- Scenario 4: Excess irrigation water is covered by treated water from the company treating and selling wastewater at a price higher than irrigation contracted-for supply (P_{EWW}) but lower than domestic contracted-for supply; no wastewater treatment on-site.
- Scenario 5: Hotel has its own wastewater treatment plant on-site (P_{TWW}) and complement excess demand from a wastewater treatment company (P_{WW}).
- Scenario 6: Hotel only uses treated wastewater on-site for irrigation (P_{TWW}) and does not buy any excess irrigation water.

Unit costs used in the model calculation are shown in Table 8.1. Table 8.2 shows an example of the model input and default parameters based on local data. These numbers can be adjusted based on location. The six scenarios were calculated for the example hotel in Sharm, considering two situations: with and without a golf course (Table 8.3).

Table 8.1 Unit prices of different types of water in Sharm (Khaled, 2008)

	Contracted-for Potable water supply (P_P)	Excess potable water supply (P_{EP}) (avg. of 3 water companies)	Treated wastewater on-site (P_{TWW}) [b]	Contracted-for wastewater [c] (P_{WW})	Excess treated wastewater supply [d] (P_{EWW})
Selling prices (US$/m^3) [a]	1.52	2.14	0.33	0.58	1.4

[a] Costs are calculated in US$ using an exchange rate of 1 US$=5.6 Egyptian pound.
[b] Cost incurred by hotel management for treating wastewater.
[c] Selling price of treated wastewater charged by Al Montaza Co.
[d] Assumed price adding cost of trucking of treated wastewater

Table 8.2 Example of input and built-in parameters for irrigation water supply optimization model for an example hotel in Sharm

Parameters	Symbol	Unit	Example values	Comments
Model Input				
No. of rooms			300	
No. of pools			2	
Pool area		m^2	800	
Av. Yearly Occupancy rate			90%	Ranging from 75% in January to 100% in July, August & September
Staff housing			yes	Staff housing exists within premises of hotel
Green area 1	A_{gr}	m^2	57,000	
Green area 2	A_{gr}	m^2	420,000	
Unit prices of different types of water		US$/m^3	Table 8.1	
Model built-in parameters				Use same values unless other information is available
Visitor consumption		m^3/c/d	0.4	
No. of visitors per room		-	2.3	
Staff consumption		m^3/c/d	0.03	
Staff housing consumption		m^3/c/d	0.25	
No. of staff per room		-	1.05	
Swimming pool consumption		m^3/m^2/d	0.01	
Factor indicating the percentage of domestic water demand ending up as sewage	f_{dw}	%	80	(Mohsen, 2007)
Optimized fraction of peak demand	f_o	%	100	Value of f_o before iteration
Specific water consumption of irrigated land	SC_{gr}	m^3/m^2/d	0.01	Maximum value calculated from CropWat 4
Average yearly temperature	T	°C	29	Average monthly temperature varies from 18 °C in January to 42 °C in August
Rainfall		mm/d	0	Effect of rainfall is ignored in the calculation due to the very low amount

Table 8.3 Example of output values for irrigation water supply optimization model for an example hotel in Sharm without a golf course (total green area (Area 1) is 57,000 m^2) and with a golf course (total green area (Area 2) is 420,000 m^2)

Parameter	**Optimized contracted-for domestic water supply**		**Fraction for Optimized contracted-for irrigation water supply**		**Optimized contracted-for irrigation water supply**		**Minimum yearly cost of domestic and irrigation water**		**Comments**
Symbol	Q_c		f_o		Q_{IC}		C_T		
Unit	m^3/month		%		m^3/month		US$/yr		
	Area 1	Area 2	Area 1	Area 2	Area 1	Area 2	Area 1	Area 2	
Scenario 1	10,084	10,084	100	100	16,432	121,084	**310,192**	**1,042,752**	Contracting for 100% of the expected peak irrigation water demand; no wastewater treatment on-site
Scenario 2	10,084	10,084	82	82	13,474	99,289	**300,412**	**970,385**	Excess irrigation water demand is covered by desalination water supply at non-contracted price; no waste water treatment on-site
Scenario 3	10,103	10,128	82	82	13,474	99,289	**300,375**	**970,226**	Excess irrigation water demand is covered by desalination water at contracted price; no waste water treatment on-site
Scenario 4	10,084	10,084	71	71	11,667	85,969	**295,970**	**937,96**	Excess irrigation water is covered by treated water from company treating and selling waste water at a price higher than irrigation contracted-for supply but lower than potable contracted-for supply; no waste water treatment on-site
Scenario 5	10,084	10,084	50	70	3,684	78,413	**267,470**	**909,461**	Hotel has its own wastewater treatment plant on-site and complements excess demand from a wastewater treatment company
Scenario 6	10,084	NA	0	NA	0	NA	**228,181**	**NA**	Hotel only uses treated wastewater on-site for irrigation and does not buy any excess irrigation water

Contracting for 100% of the peak irrigation demand is the highest cost option for a hotel. Contracting for part of the irrigation demand (in this case only 82%) resulted in cost savings of approximately 3% and 7% for a hotel without a golf course (total green area is 57,000 m^2) and with a golf course (total green area is 420,000 m^2) respectively. There is not a large difference in total cost when using desalinated water to complement irrigation water demand either at contracted-for price or at non-contracted-for price. The reason is that excess demand for irrigation occurs mainly in the summer (months of June to August) when the hotel management buys excess desalinated water at a higher price anyway. If a hotel chooses to account for the excess irrigation water demand in its initial contract for desalinated water, the only benefit is that it will secure a portion (but not all) of irrigation water and will minimize the risk of running short on water in periods of high water demand where mostly all hotels will be facing similar water shortages.

A cheaper option is to complement excess irrigation demand from the company treating and selling wastewater, if available, or from another wastewater treatment company at a higher cost (but at a cost cheaper than using desalination water) mainly due to the high demand season and the additional cost of trucking. Total savings amount to 5% and 10%, without and with a golf course, respectively. In some cases however, like in Sharm, there is only one centralized wastewater treatment plant available in the city and the hotel has no choice but to partly use desalinated water for irrigation.

Having an on-site wastewater treatment plant presents further savings on the water account amounting to 14% for both cases. But, even in this case, the presence of a golf course more than triples a hotel's total water bill. Wastewater treatment on-site covered only 8% of irrigation water requirement in case of the golf course while it covered about 60% of irrigation water requirement for the smaller green area.

Maximum available grey water is approximately 0.7 m^3 per occupied room per day and increases to 1 m^3 per occupied room per day if there is staff housing (which can increase water consumption) on the hotel premises. If a hotel only uses treated wastewater on-site (without buying any excess water) it can afford 70 (or 100) m^2 of green area for each occupied hotel room (at peak irrigation water demand of 0.01 $m^3/m^2/d$). If drought resistant vegetation is planted and the most efficient irrigation methods are adopted, this area may double. This may not be sufficient to allow for the water requirements of a golf course, but even in the case without a golf course a cost saving of 26% can be achieved.

Discussion

All wastewater should be treated and re-used for irrigation, such that expensive desalinated water (whose production can cause several negative environmental impacts) will not be used anymore for this low-value usage. The trend to construct conventional green golf courses should be halted as it requires huge amounts of blue water and in many cases the treated wastewater is not enough and needs to be complemented by desalinated water.

The least costly option for a hotel is to rely solely on treated wastewater on-site. This can be achieved by: increasing the efficiency of the irrigation system, reducing the area of high-water consuming plantation (e.g. turf grass) and/or shifting to drought resistant plants including less water-consuming grass. It should be noted that demand management of potable water may impact negatively on the availability of grey water for green areas. This has to be taken into consideration when designing the irrigation area in terms of area and type of plantation.

The efficiency of the irrigation system can be improved through careful evaluation of sprinkler head design, nozzle selection, head spacing, pipe size and pressure selection (Lazarova *et al.*, 2001). By using sophisticated on-site weather stations, weather reporting services and other resources daily irrigation requirements can be more accurately determined, resulting in significant water savings. Savings in water consumption from reducing green areas are significant. Since 1982 the United States Golf Association has invested in researches to develop new grasses that use less water and require less pesticide use. For example, a new breed of grass developed in Nebraska which replaced high water use grasses resulted in water savings of more than 50%. Another example is from the University of Georgia which has developed extremely salt-tolerant grass cultivars which can be irrigated with high-salt or brackish waters with little effect on the grass quality (Snow, 2001).

Conclusions

Hotels buy irrigation water from a water company based on a contract which stipulates a minimum daily amount of bought treated wastewater (contracted-for irrigation water supply). Water in excess of that amount can be supplied by the water company if available or complemented from expensive potable water.

Hotels have to carefully analyze their water requirements in order to determine which percentage of the hotel's peak water demand should be used in the contract in order to reduce water costs and avoid the risk of water shortage.

This paper presents a methodology to optimize the contracted-for irrigation water supply for hotels. It analyzes what the contracted-for irrigation water supply of a given hotel should be, based on the size of green irrigated area on one hand and the unit prices of the different types of water on the other hand. The objective function is to minimize the total cost of potable and non-potable water supply to a hotel. There is a need to consider cost minimization of water supply from an integrated point of view.

Contracting for the full peak irrigation demand is the highest total cost option to a hotel management. Contracting for a portion of the peak irrigation demand (calculated from the optimization model) and complementing the remaining balance from desalinated water is a cheaper option. For that reason, considering excess irrigation demand (not covered by non-potable water supply contract) while agreeing on potable water supply contract, can minimize the risk of water shortage for irrigation.

A better option still is to complement the excess irrigation demand from the company treating and selling wastewater, if available, or from another wastewater treatment company at a higher cost (but at a cost cheaper than desalinated water), mainly due to the high demand season and the additional cost of trucking. In some cases however, like in Sharm, there is only one centralized wastewater treatment plant available in the city and the hotel has no choice but to partially use desalination water for irrigation.

The least costly option for a hotel is to rely solely on treated wastewater on-site. This can be achieved by:

- increasing the efficiency of the irrigation system,
- reducing the area of high-water consuming plantation (e.g. turf grass),
- shifting to drought resistant plants including less-consuming turf grass, shrubs and trees,
- using brackish waters or even sea water to supplement irrigation water for certain types of grass.

9 Environmental impact and economic costs of brine disposal methods from RO desalination plants in arid coastal regions

Abstract

Reverse Osmosis (RO) desalination results in production of brine which needs proper disposal to avoid potential negative environmental impact. Three commonly used disposal methods are compared in this paper (sea disposal, well injection and evaporation ponds) with regard to their environmental impact and economic costs.

The chemical characteristics of reject brine from five seawater RO plants in Egypt are presented with their possible impact on marine environment and groundwater aquifers. An overview of existing regulations governing brine disposal is given. Assessment criteria to evaluate the impact of the three different brine disposal methods, chemicals concentration limits, and best practices for brine management are suggested.

The direct cost of disposing reject brine through discharge into the sea or into a well is approximately 0.05-0.06 US$/m^3 of product water. The direct cost of disposing reject brine by means of evaporation ponds is much larger and is estimated at 0.56 US$/m^3. However, the latter technology is the only method that allows for resource recovery and combined use with fish farming which may create significant benefits, making this technology competitive with well injection and direct discharge into the sea.

The indirect (environmental) costs of these three methods are uncertain. If brine is discharged into wells, the issue is whether aquifer contamination can be excluded. Since the latter is often difficult to rule out, the possible indirect cost may be high. In case of direct disposal of brine into the sea, the damage this may cause on sensitive marine ecosystems is potentially high and thus the associated indirect costs. The indirect cost of evaporation ponds has been estimated to be low.

Whereas the direct economic costs of brine disposal vary according to method and location (e.g. whether the plant is inland or coastal), the potential environmental costs are likely to be much higher. The selection of the most appropriate method thus depends on a correct estimate of the associated environmental costs.

Consistent with the above environmental considerations is the philosophy of (near) zero discharge and resource recovery. The only disposal method that would allow for resource reuse is brine disposal through evaporation ponds or brine concentrators/crystallizers, combined with economic synergies such as fish production or chemical recovery where applicable.

Keywords: Environment; evaporation pond; regulation; chemical recovery; sea discharge; well disposal.

This chapter is based on:
Lamei, A., von Münch, E., van der Zaag, P. (2009d) Environmental impact and economic costs of brine disposal methods from RO desalination plants in arid coastal regions. Accepted paper prepared for the *IDA Congress*, Dubai, UAE, 7-12 November.

Introduction

Desalination technology has contributed to alleviating water shortages in many regions, but there are nevertheless associated negative environmental impacts with this technology such as increased energy use and consequent environmental impact (Chapter 4); and potential surface or groundwater contamination resulting from improper brine management.

Brine is the highly concentrated side stream ("reject") water from the desalination process, and usually amounts to about 60-70% of the intake water. Brine contains high concentrations of salts, reaching twice the total dissolved salt concentration of the original seawater as well as chemicals used during the Reverse Osmosis (RO) desalination process (pre- and post- treatment). The traditional approach is to treat reject brine as a waste disposal problem. This trend may be changed in the future to view brine as a "saline resource" to minimize produced waste and possibly generate some revenue (Ahmed *et al.*, 2003).

In many places, regulations and standards governing brine disposal do not exist or are insufficient. This paper formulates assessment criteria to evaluate the impact of the three different brine disposal methods including concentration limits, and suggests best practices for brine management, highlighting topics for future research.

The cost of brine disposal depends on the characteristics of reject brine, level of treatment before disposal, means of disposal, volume of brine to be disposed of and nature of the disposed-to environment (Ahmed *et al.*, 2000). Current most common disposal methods include (1) direct discharge into the sea, (2) well/aquifer injection, and (3) disposal into evaporation ponds with economic synergies such as fish production and chemicals recovery. These methods differ in their environmental impact and economic costs (Glater and Cohen, 2003).

In this paper, an overview is presented on the commonly practiced disposal methods for brine from RO plants and their potential environmental impact giving examples from Sharm El Sheikh (Sharm) in Egypt. A cost comparison for the three disposal methods is included presenting capital and O&M costs from Egypt and other countries (Oman, United Arab Emirates and Israel) as well as an estimate of environmental costs.

RO desalination processes in Sharm

Data from five RO plants in Sharm was collected (4 private plants and 1 government-owned plant) and is summarized in Table 9.1 (more information on these RO plants is also presented in Chapters 2 and 3). Feedwater for these plants is seawater: beach wells for the 4 private plants and an open intake for the government-owned plant.

Data collected included feed and product water quality, chemicals used in pre- and post-treatment, method of disposal, and chemical properties of brine. Figure 9.1 shows the schematic of a general layout of the surveyed RO plants in the case study area. The figure shows where different chemicals are added along the process of RO desalination (for pre- and post-treatment). Table 9.1 explains the purpose of chemicals added during the RO treatment process for the five surveyed RO plants in Sharm.

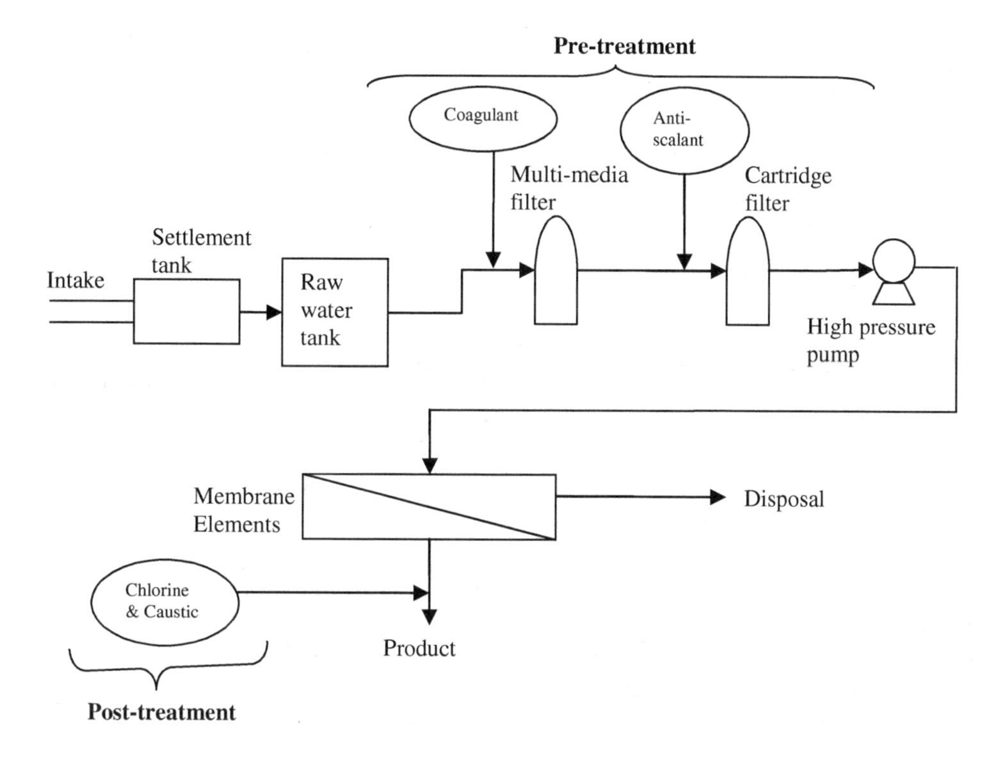

Figure 9.1 A general layout of the five surveyed RO plants in Sharm illustrating the use of chemicals, adapted from (Baker, 2006) – all five plants use the same process configuration

Table 9.1 The purpose of chemicals used during the treatment processes of the five surveyed RO plants in Sharm

Process	Chemicals (& dosage)	Purpose	Comments
Pre-treatment	Chlorine (2 ppm)	Sterilize water and bring it to a pH of 5-6	Chlorine (for pre-treatment) is only added in the government-owned plant
	Coagulants - alum or ferric chloride (from 1 to 1.5 ppm for alum and from 2 to 3 ppm for ferric chloride)	Coagulate small colloids/macromolecules to prevent them from entering the membrane	
	Ant iscalant - e.g. sodium hexametaphosphate (from 2 to 6 ppm)	Prevent deposition of sparingly soluble compounds on membrane surface	
	Sodium bisulfite (from 4 to 6 ppm)	Remove/Neutralize chlorine	Sodium bisulfite is only added in the government-owned plant
Post treatment	Caustic soda - NaOH (10 ppm)	Adjust pH for product water	Caustic soda is not added for the government-owned RO plant
	Chlorine (1 ppm)	Disinfect product water	

In the four private RO plants surveyed, chemical analysis of the reject brine was only performed regularly for one plant (Sinai RO plant), and was only done once on request (for this research) for the remaining three. For the government-owned plant, no brine composition tests were performed at all. Disposal to sea (government-owned plant) or to wells (private plants) was done regardless of the brine composition.

Analysis of the chemical composition of brine is important as toxic chemicals generated during the RO desalination process may pose a potential risk for contamination of aquifers if a leakage occurs from a disposal well or an evaporation pond. Also, brine characteristics such as concentration, discharge rate, and outlet pressure impact well performance (injectivity efficiency). For direct seawater discharge, brine characteristics and design of the disposal pipeline system influence the extent of damage to the marine environment (Table 9.3) (Einav and Lokiec, 2003).

Table 9.2 presents characteristics of reject brine from the four private surveyed RO plants in Sharm (Egypt). Table 9.2 shows that manganese in the brine was high for the sample from the Sinai plant (4.2 mg/l compared to 0.1 and 0.3 mg/l from the other plants) as well as calcium (5,854 mg/l compared to an average of 934 mg/l from the other plants). This is due to high levels of manganese and calcium detected in raw water in that particular location. The iron content (Fe) is about twice as high in the concentrate as in the Red seawater. This can be attributed on one hand to the concentration effect of the RO desalination and on the other hand to the discharge of the coagulant and the media filter backwash with the concentrate. On another point, although the used anti-scalant contained phosphate, no phosphate measurement was done for any of the plants.

Table 9.2 Characteristics of reject brine from four private-owned seawater RO plants in Sharm (Egypt), only one sample was available from each plant (Girgis, 2008), and Red Sea water (Khaled, 2008)

Parameters	Red Sea Water	Coral Sea (Sharm)	Sinai (Sharm)	Millennium (Sharm)	Magic life (Sharm)
Capacity m^3/d	NA	500	6,500	600	1,000
Disposal method	NA			well	
Temperature, °C	23	27.5	27.5	27.5	27.5
pH	8.3	7.6	7.2	7.6	7.7
Electrical conductivity, µS /cm	67,000	80,500	81,700	58,000	77,500
Ca, mg/l	440	1,003	5,854	802	993
Mg, mg/l	1,520	2,214	2,657	1,429	1,936
Mn, mg/l	NR	NR	4.2	0.1	0.3
Na, mg/l	12,700	13,400	16,200	8,000	12,600
HCO_3, mg/l	NR	167	126	130	NR
SO_4, mg/l	3,055	4,850	2,860	3,350	4,250
Cl, mg/l	22,840	NR	36,250	NR	NR
TDS, mg/l	40,000	54,650	55,380	39,376	52,614
Total hardness, mg/l	7,400	10,216	23,820	6,977	9,220
Free Cl_2, mg/l	NR	0.1	0.1	0	0.1
SiO_2, mg/l	0.35	NR	NR	NR	NR
Cu, mg/l	Nil	NR	NR	NR	NR
Fe, mg/l	0.06	0.1	0.2	0.1	0.1
Ni, mg/l	NR	NR	NR	NR	NR

NA = not applicable, NR = not recorded

Table 9.3 lists the main chemical parameters (from Table 9.2) of brine and their impact on the marine environment or groundwater aquifers.

Table 9.3 Impact of the main chemical characteristics of brine on marine environment and groundwater aquifers

Parameter	Potential Impact
Calcium & magnesium	Increase the salinity of water it comes in contact with in the disposal aquifer. The problem is more severe if the brine leaks into a potable groundwater aquifer (Semiat, 2000)
Salinity	Most marine organisms will not survive in case of a high increase in salinity; also causes seawater turbidity (Al Agha and Mortaja, 2005; Einav *et al.*, 2002)
Total dissolved solids (TDS)	High TDS levels can cause well plugging and decrease injectivity efficiency of a well (Glater and Cohen, 2003; Vedavyasan, 2001)
Free Cl_2	Toxic and affect biological and enzymatic process of living organisms (Einav *et al.*, 2002)
Temperature	Increased temperature reduces dissolved oxygen which is necessary for the respiration of marine organisms (Einav *et al.*, 2002)
Iron	Additives like iron are mainly dark in color, they cause turbidity to seawater preventing penetration of light and therefore disrupting photosynthesis (Svensson, 2005)
Copper, titanium & Nickel	From corrosion of equipment, toxic towards marine organisms and humans (Einav *et al.*, 2002)

Potential environmental impact of different brine disposal or treatment methods

Current knowledge of quantifying and evaluating the impact of brine disposal on marine life or shallow aquifers is limited. This section provides an overview of the environmental impacts of the three methods of brine disposal considered in this paper. This overview is based on a limited number of published field experiences.

Direct sea discharge
Direct sea discharge has been the most common method for brine disposal used in Egypt (and worldwide as well) for coastal RO plants built to meet the increasing water demand of the tourism industry.

The Red Sea is characterized by the presence of important coral formations which lead to a high diversity of marine life. It is usually calm slowing down natural mixing processes. From the surveyed plants, only the government-owned plant discharges brine directly into the sea. Whether or not the brine disposal from this plant has a measurable impact on the surrounding coral sea life has not yet been researched. It is clear from Table 9.2, however, that reject brine is more acidic than the Red Sea water with higher temperature and higher concentration of chlorides, TDS and Total Hardness (except for the Millennium plant).

Direct sea disposal is preferred for sandy bottoms without flora. When concentrate with a higher density is disposed into water of lower salinity (lower density), the concentrate tends to sink introducing problems for the marine environment. If discharge to sensitive marine beds is inevitable, brine must be diluted to acceptable limits as suggested by the competent Environmental Authority depending on the nature of marine organisms and potential impact of discharge (Latorre, 2005).

In all cases, natural dispersion, mixing and hence dilution will occur depending on the discharge pipe's location, waves, tides, currents and water depth. Without proper dilution the brine plume may extend for hundreds of meters beyond the mixing zone harming the ecosystem along the way. Mixing zones are defined area limits within the receiving waters where the law allows surface water to exceed water quality standards due to the existence of point source disposal. When natural dilution is not sufficient to properly diffuse the concentrate, then desalination plants use artificial dilution methods: efficient blending prior to surface disposal or diffusers (Younos, 2005).

In Australia, a dilution of up to 45 fold was applied to reduce brine salinity into that of the original seawater. As a result of the dilution, total dissolved solids (TDS), total suspended solids (TSS), iron, sulphates, and chlorides were almost equal to original seawater (Strategan, 2004). Less-toxic chemicals that do not harm the environment will probably replace the added chemicals (for pre- and post-treatment) in the future, such as organic anti-scalants (Semiat, 2000). Optimized dosing of chemicals through correct monitoring of the desalination process will help minimize the environmental impact.

Well disposal

Another disposal method for brine from RO plants practiced in Egypt is well disposal to an aquifer. The disposal well (about 60 m depth) is drilled about 50 m away from the feed well (30 m depth) regardless of the suitability of geological formation. There were several incidents in Sharm where interference between feed and reject water occurred due to close proximity of feed and reject wells (Girgis, 2008).

This interference leads to increased salinity of feed water negatively impacting the performance of the RO desalination process (Hafez and El Manharawy, 2002). Altering the chemical properties of the original water in the aquifer can impact the usage of the aquifer for any other application especially at larger temporal and spatial scales (Delleur, 2007).

In Sharm the aquifers into which reject brine is injected, discharge directly into the sea. This could have a negative environmental impact similar to direct sea discharge. The highly saline plume in the aquifer migrates slowly to the coastal area with little mixing, once the plume has reached the sea it may be a long-lasting source of high salinity. Careful assessment of geological conditions in order to determine depth and location of porous aquifer reservoir should take place before injection of brine into the aquifer takes place (Glater and Cohen, 2003). Usually the possibility to monitor the discharge in the disposal aquifer is inadequate resulting in uncertainty of environmental impacts (Svensson, 2005).

Evaporation ponds for treatment and disposal

Most desalination plants in Egypt are located close to the coastline and direct sea discharge is the most common method except for highly sensitive marine areas. Evaporation ponds as a brine treatment and disposal method is currently not commonly used in Egypt, but this method has been applied successfully in places such as Israel, Australia, China, India, some of the Arabian Gulf countries and Chile.

Evaporation ponds are designed to concentrate the received brine and reduce its volume through evaporation. The disposal of the dry end product is usually done to approved waste disposal sites. Evaporation ponds are a relatively cheap means of disposal in areas with high evaporation rates, low rainfall and low cost of land (Svensson, 2005). The cost of evaporation ponds is however very sensitive to the value of land, due to the large surface area required. Evaporation ponds can also be used for fish production and chemicals recovery to increase the economic benefits of the system. Brine contains different chemicals which could be recovered for sale, e.g. magnesium hydroxide, sodium chloride, calcium carbonate, sodium sulphate and calcium chloride (Ahmed *et al.*, 2003) .

In Eilat, Israel, an RO plant, which has been in operation since 1998, not only produces RO desalination water but also sodium chloride: brine from the RO plant is pumped into a series of evaporation ponds and then to the salt processing factory (Ravizky and Nadav, 2007). Occasionally, the pond has to be emptied to remove some of the salts that have crystallized. The disposal of a dry end product is much more economically feasible than liquid disposal; usually the disposal is done to approved salt deposit sites (Svensson, 2005) or used for salt recovery. One potential problem for evaporation ponds is leakage which may result in groundwater contamination. To prevent this, most of the current installations are lined with polymeric sheets (Glater and Cohen, 2003).

Environmental regulations governing brine disposal
In many places, regulations and standards governing brine disposal do not exist or are insufficient. This section gives an overview of the existing regulations for brine disposal present in the USA, Australia and Egypt and suggests assessment criteria to evaluate the three different brine disposal methods mentioned in this research: sea discharge, well disposal and evaporation ponds (Table 9.7). In case of discharge to surface water, the United States Environmental Protection Agency (USEPA) requires a demonstration of acceptable brine chemistry such as pH, total suspended and dissolved solids, and individual chemicals. The limits are based upon the nature and use of the receiving water body, and human and aquatic toxicity studies. Effluent toxicity tests are performed where different organisms common to the receiving water are exposed to brine and survival and growth is monitored. Periodic testing and monitoring of specific chemicals concentrations is required (Svensson, 2005).

Literature does not specify a salinity limit above which definite damage is caused to marine population. However, (Einav and Lokiec, 2003) suggested that the salinity range should be minimized to the salt concentration of the original seawater, i.e. brine has to be diluted. The Australian Environmental Protection Agency sets limits for salinity levels, temperature, and dissolved oxygen. In Egypt, there are no national regulations for brine disposal. However, there are regional regulations prohibiting direct sea discharge in marine sensitive areas (e.g. Red Sea).

In the USA, permission for injection of brine into wells must be granted. Requirements include monitoring well integrity and water quality of nearby monitoring wells. Any water with salinity below 10,000 mg/l is considered a potential source of drinking water. Regulations in Florida prohibit injection in groundwater aquifers which can be a possible drinking water source (Kimes, 1995). Brine injection wells are currently classified by the USEPA as class V wells (i.e. wells for non-hazardous waste) for which there are no restrictions on well location or concentrate concentration. However, most states require a hydrogeological study to confirm that no contamination of aquifers will occur (USCongress, 1988). For permission to use evaporation ponds, the USA typically requires pond integrity to be monitored.

The most suitable disposal method from an environmental and economic perspective has to be evaluated site specifically. An effort has been made in this paper to compile assessment criteria to aid in the selection and evaluation of brine disposal methods. Limits or best practices to evaluate the assessment criteria are suggested as well as areas where future research is still needed.

Economic analysis including environmental costs for different brine treatment and disposal methods

Any project includes both direct and indirect costs. In most cases, direct costs are the only factors considered when project costs are being estimated. In general, direct costs are easier to estimate compared with indirect costs, mainly due to the larger uncertainties involved in the latter (Semiat, 2000; USEPA, 1992). Moreover, and importantly, project owners often do not bear the indirect costs and therefore have little interest in estimating them.

In this section, an attempt is made to calculate both direct and indirect costs for the three disposal methods discussed in this paper. The methodology used is similar to (Gordan, 2001) in which the author incorporates environmental costs in the economic analysis of different water supply options. Desalination was one of the analyzed options. However, the author only described the impact of brine disposal qualitatively reasoning that the magnitude of impact on the environment is uncertain and the indirect cost (environmental) estimates are not available in literature (Gordan, 2001).

Direct costs are investment (capital) and operational costs while indirect costs (or environmental costs) are the external and user costs. The following equation is used to calculate the total cost of a project (adapted from (Pearce and Markandya, 1989)):

$$TC = DC + EC + UC \qquad \text{Equation 9.1}$$

where TC is opportunity cost measured in economic terms (US$/m^3); DC is direct cost (US$/m^3); EC is external cost (US$/m^3) and UC is user cost (US$/m^3).

Direct costs (and benefits) of brine disposal

According to Ahmed *et al.* (2001), the capital cost of a brine disposal installation ranges from 5 to 33% of the total capital cost of the RO desalination plant. The capital cost of brine disposal in wells is 4 to 5% of the total capital cost of an RO desalination plant. The cost of well disposal includes the disposal pump station and pipeline cost in addition to the construction cost of wells. Unit capital cost ranges from 61 to 84 US$/m^3/d (2001 costs) of product water for capacities ranging from 250 to 4,800 m^3/d and reject brine volumes ranging from 306 to 11,200 m^3/d (Hafez and El Manharawy, 2002).

The capital cost of surface discharge into the sea includes the disposal pump station and the disposal pipeline cost. Using the data presented in Hafez and El Manharawy (2002), the capital cost of direct sea disposal was calculated by subtracting the cost of well construction. Accordingly, the capital cost of direct sea disposal is about 2% of total capital cost of an RO desalination plant and ranges from 33 to 46 US$/m^3/d of product water.

The cost of evaporation ponds includes the pumping and pipeline disposal system in addition to the cost of land acquisition (a major cost component), excavation, earth work, lining material and the cost of solid salt disposal. Ahmed *et al.* (2001) reports capital costs for five evaporation ponds in Oman and United Arab Emirates (4 plants of 100 m^3/d capacity and one plant of 1000 m^3/d capacity). Unit capital cost for the 1000 m^3/d plant is 384 US$/m^3/d while cost ranges from 467 to 1,534 US$/m^3/d of product water for the four 100 m^3/d plants. The large variation in cost for the 100 m^3/d plants is due to cost of land, remoteness of plant location, and availability of local construction material and labour (Ahmed *et al.*, 2001).

The following equation is used to estimate the area required for an evaporation pond as a function of evaporation rate and volume of brine (Ahmed *et al.*, 2000).

$$A = \frac{Q_r f}{e_v} \qquad \text{Equation 9.2}$$

where A is open surface area of evaporation pond (m^2); Q_r is volume of reject brine (m^3/d); e_v is evaporation rate (m/d); and f is a safety factor. e_v is taken as 10 mm/d and f as 1. Rainfall is assumed negligible.

Note that enhanced evaporation mechanisms such as spraying of brine, creating turbulence in the pond and creating airflow over the ponds, can accelerate evaporation and reduce the area of the evaporation pond.

The O&M costs for the three disposal methods include maintenance and repair costs. It ranges from 0.01 to 0.03 US$/$m^3$ for evaporation ponds and surface and well disposal respectively.

Table 9.4 summarizes capital and O&M costs (direct costs) for the three disposal methods for an RO plant of capacity 250 m^3/d producing 305 m^3/d of brine. The direct cost of evaporation ponds (0.55 US$/$m^3$) is one order of magnitude larger than that of direct discharge of brine into the sea (0.05 US$/$m^3$) or into wells (0.06 US$/m^3). Note that Glueckstern and Priel (1996) quote an even higher unit cost of brine disposal using evaporation ponds, namely US$ 0.96/m^3 (data of a desalination plant located in the Negev region, Israel, producing 384 m^3/d of brine, cost adjusted for year 2001).

Table 9.4 Capital and O&M costs (direct costs) and benefits of the three methods of brine disposal (costs adjusted for 2001)

	Capital Cost (US$)	Unit capital cost (US$/$m^3$/d)	Amortized capital cost[b] (US$/$m^3$)	O&M (US$/$m^3$)	Total direct cost (US$/$m^3$)	Total benefits (US$/$m^3$)	Total direct cost after deducting benefits (US$/$m^3$)
Surface sea discharge	11,535 [a]	46	0.02	0.03[a]	0.05	0.00	0.05
Wells	21,119 [a]	84	0.03	0.03[a]	0.06	0.00	0.06
Evaporation ponds	337,458[c]	1,349	0.55	0.01[d]	0.56	0.55	0.01

[a] (Hafez and El Manharawy, 2002)

[b] Using discount rate of 8% and lifetime of 10 years

[c] using empirical equation developed using data from Ahmed *et al.* (2001): $2.8Q_r$+252.5, where Q_r is vol. of reject brine (m^3/d)

[d] O&M cost for evaporation ponds is 2% from annual capital cost (USDI, 2002)

There are also potential benefits that can be gained from reject brine, namely through the recovery of useable chemicals. This is only possible with methods where the brine is concentrated (i.e. evaporation ponds) and not with methods that aim to disperse the brine into the environment (i.e. well disposal, surface sea water disposal). No data was available on the potential net economic benefits of chemicals recovery from reject brine (see further the discussion section below). Moreover, additional benefits may be gained by combining evaporation ponds with fish farming.

The potential benefits of fish farming in evaporation ponds may be large: an evaporation pond with a capacity to receive 305 m^3/d of reject brine should have a surface area of approximately 30,500 m^2 (Equation 9.2). If such a pond has a capacity to produce two tilapia fish per m^2 per year, with an expected net profit of at least US$0.8/fish (Svensson, 2005), this would translate into an additional benefit of at least 0.55 US$ per m^3 of product water.

Indirect costs of brine disposal

External costs are positive or negative attributes or effects of a good/service or its production not reflected in the price of the production/service, e.g. human health cost associated with particles matter emissions from a coal power plant. External cost can involve administration costs, and regulatory compliance costs: permitting, monitoring, liability costs, penalties, fines, personal injury, property damage, and natural resources damage clean up costs through a superfund or a corrective action (Young, 2005).

External costs (e.g. liabilities) are estimated based on the damage cost method measuring the resource costs brought on by an environmental change (Young, 2005). The net present value of the cost was estimated based on actual case studies where law suits where filed against a polluter to a water resource (not a desalination plant) and a fine (damage cost C_D) had to be paid (Equation 9.4). Evaporation ponds were assumed as having minimum probability of groundwater pollution due to the fact that most current facilities install liners.

User cost is estimated as the present value cost of replacing an environmental asset or the lost revenue at some future point due to depletion (as an example) assuming that the direct cost/ revenue from the existing technology remains constant. The following scenarios for potential impact from the three brine disposal methods and the consequent user costs are as follows (Table 9.5):

- Surface discharge: if brine is disposed without dilution into a sensitive marine environment, ecological degradation is expected, and over the time tourism is likely to get impacted. Revenue from tourism will decrease (Equations 9.3 and 9.4).
- Well disposal: if the water in the disposal aquifer gets in contact with the water supply aquifer for desalination, gradual increase in salinity will take place rendering the water supply aquifer unsuitable for desalination over time. The problem is more severe if the disposed brine leaks into a potable groundwater aquifer. As a result, a shift to another water supply option may take place, e.g. long-distance piping (Equation 9.4).
- Evaporation ponds: The user cost of evaporation ponds is estimated as zero, since the land cost have been included in the calculations of the direct costs, and the price of the land is assumed to include its opportunity cost, and hence the alternative uses of the land.

For surface disposal, in order to obtain the lost revenue from decreased tourism, the following equation is used (adapted from (Gordan, 2001)):

$$C_{LR} = [TR / l \times a_b] / Q \qquad \text{Equation 9.3}$$

where C_{LR} is lost revenue per cubic meter of desalinated water for loss of beach access (US$/$m^3$); TR is the total expected yearly revenue from an activity (US$/yr); l is the length of shoreline (km); a_b is beach access lost due to water quality degradation (km); Q is total water consumption of the entire beach area (m^3/yr).

The net present value of external and user cost is obtained using the following equation (adapted from (OECD, 1994)):

$$C = C_{D,LR,R} / (1+i)^{y}$$ Equation 9.4

where C is net present value of user or external cost (US$/m^3); $C_{D,LR,R}$ is cost of damage, or lost revenue (Equation 9.3) or cost of replacement technology which is P_R-C_E (where P_R is price of replacement technology (US$/m^3) and C_E is direct cost of existing technology (US$/m^3)); i is discount rate (%); y is number of years until lost revenue or damage takes place or backup technology replaces existing technology (yr). The value of i and y are taken as 8% and 20 years, respectively.

Table 9.5 estimates environmental costs (indirect costs) associated with the three brine disposal methods. The indicated values are indicative costs of possible likely environmental damage based on analogies from similar domains. These values are approximations given the many uncertainties involved.

Table 9.5 Environmental costs (indirect costs) associated with brine disposal (costs adjusted for 2001)

Disposal method	Type of potential impact	Type of indirect cost	Cost US$/m^3	Comments
Surface sea discharge	Lost revenue from tourism	User cost	0-0.2[a]	Lower range for long pipes deep into the sea and sandy bottoms, higher range for outlets near shore line and abundance of marine life
	Surface water pollution	External cost	4.13[b]	ocean clean up, in case of coral reefs damage, situation is irreversible
Wells	Excess cost due to shifting to another water supply resource	User cost	0.19[c]	Extra Cost of potable water transportation by pipelines
	Groundwater pollution/ increased salinity	External cost	0.34[d]	Law settlement for pumping & removing contaminates from potable groundwater aquifer (treating 13,000 m^3/d)
Evaporation ponds		User costs	0	Cost of land is included in direct costs and is assumed to include to some extent the alternative uses of land
	Groundwater pollution	External cost	0	Most of the current installations are lined with polymeric sheets

[a] Using Equations 9.3 and 9.4, where total revenue from tourism in Sharm for year 2001 (TR) is 1,030 Million US$ (WTO, 2007); Sharm beach length (l) is 32 km; shoreline that can be impacted by disposal (a_b) is 0.8 km (i.e. maximum permissible limits for a mixing zone where water quality can exceed water quality standards) (Younos, 2005); and Q is 73,800 m^3/d

[b] Using Equation 9.3; total cost of ocean clean up was 105 Million US$, and water treated volume was assumed as 13,000 m^3/d of seawater (Unknown, 2008)

[c] Using Equation 9.3 assuming desalination cost (C_E) of 1 US$/m^3, pipeline cost for length of 368 km and capacity $\leq$ 2,000 m^3/d (P_R) is 1.9 US$/m^3 (Chapter 3)

[d] Using Equation 9.3, total settlement was 8.35 million US$ and water treated volume was 13,000 m^3/d of groundwater (Directory, 2005)

Discussion

Whereas the direct economic costs of brine disposal vary according to method and context (e.g. whether the plant is inland or coastal), the potential environmental costs (i.e. indirect costs) are likely to be significant (Gordan, 2001). This implies that excluding the indirect costs in a comparative economic analysis may lead to considerable biases; hence the importance of estimating these indirect costs.

The main problem with measuring the environmental impact is the large uncertainties involved, depending on brine disposal method, type of ecosystem absorbing the disposed brine, and the interaction between the brine and the receiving ecosystem. However, when expected damage from a certain disposal method is irreversible or beyond the human timescale in order to recover, the precautionary principle applies and this disposal method should not be adopted (Loucks and Van Beek, 2005). This is the case for damage to coral reefs from surface discharge or damage to groundwater aquifers (which can be used for potable water supply now or in the future) where the storage residence time is long and consequently the negative impact on the water quality will be long-lasting.

Though it is difficult to estimate the cost of environmental impact in economic terms, they are likely to be significantly larger than the direct cost, especially with respect to direct sea disposal and well injection. If only direct costs are taken into account, and if the desalination plant is located near the sea, and it is not feasible to use evaporation ponds productively (fish farming/chemical recovery), than direct discharge into the sea would be selected. If however the indirect costs are adequately accounted for and the environmental impacts are appropriately taken into account, disposal wells are a better option for coastal RO plants while evaporation ponds can be more suited for inland plants, especially where the value of land is likely to remain low.

Apart from environmental costs, a (near) zero discharge goal and reuse and recovery philosophy may be included in the analysis (Crites *et al.*, 2000). A philosophy of reuse and resource recovery would favour the evaporation pond technology, since both other options do not have any opportunity of reuse or resource recovery. Dumping rather than treating waste transfers costs to others distant in time and space (Young, 2005).

Recovery and sale of chemicals can improve the economic feasibility of brine disposal through evaporation ponds. Producing the salts or chemicals to a purity which make them marketable requires additional processes that involve multiple evaporation and/or cooling, supplemented by chemical and mineral processing (Ahmed *et al.*, 2003).

Table 9.6 presents a list of chemicals that can be recovered from reject brine and their current selling price in Egypt. In Ahmed *et al.* (2003), the expected quantities of salts produced from an amount of 75,000 m^3/yr of reject brine were reported (these values vary from plant to plant depending on treatment process and chemical composition of the brine).

- Gypsum: 350 tons/yr
- Sodium chloride: 1,000 tons/yr
- Magnesium hydroxide: 75 tons/yr
- Calcium chloride: 240 tons/yr

Table 9.6 List of chemicals that can be recovered from reject brine of an RO desalination plant and current market selling price in Egypt (adapted from (Ahmed *et al.*, 2003))

Product name	Chemical composition	Current market price in Egypt (2008) (US$/t)	Potential applications
Gypsum	$CaSO_4.2H_2O+Mg(OH)_2$	301	Soil remediation Fertilizer additive
Magnesium hydroxide	$Mg(OH)_2$	NA	Wastewater treatment Agriculture Cattle feedstock additive
Sodium chloride	NaCl	301	Food processing Agriculture
Calcium carbonate	$CaCO_3$	257	High value paper coating pigment Filler in plastics paint, ink and sealant production
Sodium sulphate	Na_2SO_4	792	Pulp & paper industries
Calcium chloride	$CaCl_2$	1,075	Road base stabilization Soil remediation Dust suppression

Conclusion

Desalination technology has contributed to alleviating water shortages in many regions, but there are nevertheless associated negative environmental impacts, including high energy consumption and the potential surface and groundwater contamination caused by reject brine disposal. Chemicals used in pre- and post-treatment within the RO desalination process enrich reject brine with ions. The common notion is to dispose of brine as a waste product. However, brine could also be seen as a potential source of useful salts and chemicals.

In many places regulations and standards governing brine disposal do not exist or are insufficient. The paper provides an overview of some of the existing regulations for brine disposal present in the USA, Australia and Egypt and suggests assessment criteria to evaluate the three brine disposal methods considered in this paper. Currently there are no national regulations for brine disposal in Egypt, although there are some local regulations in sensitive areas prohibiting direct sea discharge in marine sensitive area (e.g. Red Sea). The Egyptian Environmental Protection Agency is beginning to acknowledge the problem as desalination activities are growing.

The cost of reject brine disposal depends on the characteristics of the brine, level of treatment before disposal, means of disposal, volume of brine to be disposed of and nature of the disposal environment (Ahmed *et al.*, 2000). Current main disposal methods include direct sea discharge, (deep) well injection and evaporation ponds. The paper estimated the direct and indirect (including environmental) costs of the disposal of reject brine through these three different methods. The direct cost of disposing reject brine through discharge into the sea or into a well is approximately 0.05-0.06 US$/m^3 of product water. The direct cost of disposing reject brine by means of evaporation ponds is much larger and is estimated at 0.56 US$/m^3. However, the latter technology is the only method that allows for resource recovery and combined use with fish farming which may create significant benefits, making this technology competitive with well disposal and direct sea discharge.

The indirect cost of these three methods is uncertain and largely depends on whether environmental damages will or will not occur. If brine is discharged into wells, the issue is whether aquifer contamination can be excluded or not. This depends on the local geology and the non-connectivity of the disposal well with the source water aquifer. Since the latter is often difficult to rule out, the possible indirect cost may be high. In case of direct disposal of brine into the sea, the question is what damage this may cause on sensitive marine ecosystems. This damage is potentially high and thus the associated indirect costs. The indirect cost of evaporation ponds has been estimated to be low.

The following are recommendations for brine management in order to reduce potential negative environmental impact:

1. Whereas the direct economic costs of brine disposal vary according to method and location (e.g. whether the plant is inland or coastal), the potential environmental costs are likely to be much higher. The selection of the most appropriate method thus depends on a correct estimation of the associated environmental costs.
2. If only direct costs are taken into account, and it is not feasible to use evaporation ponds productively (fish farming/chemical recovery), and if the desalination plant is located near the sea, than surface discharge in the sea is the cheapest option. If, however, the indirect costs are adequately accounted for, evaporation ponds may be a more suitable option, especially where the value of land is (expected to remain) low.
3. Consistent with the above environmental considerations is the philosophy of (near) zero discharge and resource recovery. The only disposal method that would allow for resource reuse is brine disposal through evaporation ponds or brine concentrators/crystallizers, combined with economic synergies such as fish production or chemical recovery where applicable.

To date the problems associated with disposal of brine has not been significant enough to overrule building a desalination plant. However, with increasingly stringent environmental regulations, disposal of brine could become a constraint in the feasibility of future RO plants.

Acknowledgements
We are grateful to Mr. Zaki Girgis and Mr. Haytham Ahmed, Ridgewood Egypt; and Mr. Tarek Waly, UNESCO-IHE, for kindly providing valuable data for the research.

Table 9.7 Suggested criteria to assess impact of brine disposal

Disposal method	Assessment criteria	Suggested limits/standards/practices	Area of research	References
Sea discharge	Impact of chemicals used in pre- and post- treatment on marine life	-Use low toxicity and rapid breakdown chemicals -Perform toxicological testing to determine impact of different chemicals on marine ecosystem	-Monitoring studies are needed to assess impact of brine sea discharge on marine ecosystem -Research is needed to fully understand impact of chemicals (stability, residence time and eco-toxicity) and set acceptable concentration limits	(Shahatto, 2003; Strategan, 2004)
	Chlorine	Not allowed especially in water with a hydrocarbon content (e.g. Arabian Gulf) due to possible formation of carcinogenic by products		(Svensson, 2005)
	Dissolved oxygen	90%		(Strategan, 2004)
	Temperature ºC	±0.5 in dry conditions		
	Salinity (ppm)	±800		
	Dilution factor	-salinity range should be minimized to the salt concentration of the original seawater - use of diffusers with sufficient ports, diameter and spacing to ensure required dilution - Monitoring of dilution process & parameters such as TDS, dissolved oxygen, temperature and salinity		(Einav and Lokiec, 2003; Strategan, 2004)
Well disposal	Presence of an aquifer with no connectivity to original source water	- Detailed assessment determining depth and location of a suitable porous aquifer reservoir , nature of subterranean strata to determine permeability and solution confinement capability		(Glater and Cohen, 2003)
	Injection well performance	-Frequent measurement of total suspended solids (TSS)	Limit to acceptable TSS can be derived from models relating TSS with injection rate & pressure, and porosity and permeability of strata	(Glater and Cohen, 2003)
	Demonstration of acceptable brine chemistry such as pH, total suspended and dissolved solids, individual chemicals	- Brine conditioning & filtration required prior to injection - Monitoring of water quality in nearby wells		(Glater and Cohen, 2003)
Evaporation ponds	Design (surface area, depth and freeboard)	-Area directly proportional to volume of reject brine & inversely proportional to evaporation rate -Freeboard is a function of rainfall intensity and wind velocity	Optimized evaporation ponds design can be done through experimental circulation of typical brine solution through small evaporation ponds	(Glater and Cohen, 2003)
	Leakage from ponds	Double lining is recommended with special attention to sealing joints		
	Monitoring of pond integrity	Leakage sensing probes between layers of lining		

10 Conclusions and recommendations

General

Water scarcity aggravates in coastal zones which are characterized by high population density, intense economic activity and tourism, and consequently heavy seasonal water demand. The usual way to tackle water demand is through conventional surface and ground water abstraction. However, due to increasing limitation in water resources, a shift is taking place towards integrated water resources management (IWRM).

IWRM adopts a holistic approach to optimize water usage. IWRM has to take into account the following four dimensions: water resources, water users, spatial and temporal scales. It critically assesses supply options, including developing alternative water resources, e.g. desalination (removal of dissolved minerals including salts) and reclaimed wastewater. It also seeks to increase the management efficiency of conventional resources and schemes. Finally, and importantly, an IWRM approach will also consider demand management options. IWRM projects should be sustainable and fulfil public interest: maintenance of environmental quality, financial sustainability, good governance, institutional capacities and social equity.

However, the development, decision making process and implementation of projects conforming to the IWRM approach are complex due to the different sectors involved, typically water, environment, energy and agriculture.

Egypt is among those countries which are very vulnerable with regard to water resources. It is located in a belt of extreme aridity being the country furthest downstream in the Nile basin. The pressure of water scarcity, for regions located along the length of the Nile, is already mounting. A more disadvantaged region is the Red Sea coastal area, where fresh water is not always available.

The scope of this study was to develop a technical-economic modeling tool to aid decision makers, both public and private sector investors, in the design and assessment of financial and environmental sustainable integrated water resources management projects for tourism-dominated arid coastal regions. Designed projects are to satisfy varying temporal and spatial demands and to find sustainable solutions, which may include a higher level of wastewater reuse.

The evaluation of integrated water resources management (IWRM) projects should be based on a cost benefit analysis and should also consider environmental impact. The decision maker using the technical-economic model can be a hotel, a water company, or a water management company (i.e. municipality, city, town, and governorate). The water management company is assumed to perform integrated water management including: supply of potable water, collection and treatment of wastewater, and reuse of treated wastewater for irrigation water supply.

The research problem emerging from the first chapter was how decision makers plan for sustainable integrated water resources management projects taking into consideration the complexity of the projects due to the involvement of several factors: water, environment, energy and agriculture. Chapter 2 presented information on the case study region: Sharm El Sheikh (Sharm) in Egypt regarding: water demand, water supply options in the regions, current water resources management practices, and shortfalls including increasing water shortages and price rises as well as environmental degradation. Simple strategies are outlined which could be undertaken to improve the situation.

Chapter 3 presented a cost comparison for two options to supply water of drinking water quality: Desalination with the reverse osmosis technology, or long-distance water piping from the Nile. Chapter 4 analysed unit production costs and energy costs for 21 reverse osmosis (RO) desalination plants in the region. An equation was derived to estimate the unit production costs of RO desalination plants as a function of plant capacity, price of energy and specific energy consumption. This equation was used to calculate unit production costs for desalinated water using photovoltaic (PV) solar energy based on current and future PV module prices. Unit production costs of desalination plants using solar energy were compared with conventionally generated electricity considering different prices for electricity.

Chapter 5 presented an analysis of the water demand by the tourism industry in arid coastal regions, i.e. hotels and related services. A model was developed to calculate time-variant water demand by the tourism industry. The model also considered the impact of introducing demand management to reduce water consumption in hotel facilities. Chapter 6 presented the setup of potable water supply to hotels in the case study region. Private water companies supply these hotels with their domestic water needs. There is normally a contractual agreement stating a minimum requirement that has to be supplied by the water company and that the hotel management has to pay for regardless of its actual consumption ("contracted-for water supply"). An optimization model to determine what value a hotel should choose for its contracted-for potable water supply in order to minimize its total annual water costs as a function of occupancy rate and hotel characteristics was developed. Chapter 7 presented a model using Excel macros to perform dynamic optimization with the objective function to maximize present value of total net benefits over the lifetime of an RO desalination plant. The aim of the dynamic optimization is to solve for capacity expansion. The model can be used to test different scenarios to capture time-variant tourism demand and price uncertainties on investment decisions.

Chapter 8 described a model to optimize contracted-for irrigation water supply with the objective function to minimize total water cost to hotels. Chapter 9 discussed three commonly used brine disposal methods which are compared (sea disposal, well injection and evaporation ponds) with regard to their environmental impact and economic costs. Assessment criteria to evaluate the impact of the three different brine disposal methods, chemicals concentration limits, and best practices for brine management were suggested. Estimates of the direct and indirect costs of the three disposal methods were presented.

The first part of this chapter presents the conclusions regarding the research problems and the associated research questions. Following this, the implications of this research are briefly discussed. Subsequently the limitations of the research are discussed. Finally opportunities for further research are outlined.

Research Questions

In this section, each of the research questions is stated together with a summary of the findings from this research.

- *What is the current setup for water resources management in tourism-dominated arid coastal regions, with the example of Sharm El Sheikh? What are the shortfalls and how can they be addressed?*

The city of Sharm is a typical example of a tourist city located in an extremely arid environment with a haphazard approach to water resources management. Water supply to Sharm is mainly from RO desalination complemented by groundwater from Al Tor which is transported by tankers or long-distance pipelines as well as from treated domestic wastewater (for landscape irrigation). Desalinated water is provided by two government-owned RO plants, two centralised privately-owned RO plants and by about 50 decentralised small RO plants (Khaled, 2008). The government-owned RO plants are selling water at a very low subsidized price to the local residents while the two centralised private RO plants (owned by two different companies) could control the market and raise prices considerably especially in periods of high water demand (selling price ranges from 1.6 to 2.5 US$/m^3). All of these RO plants cause environmental problems due to high energy consumption per m^3 of water produced and the impact of uncontrolled brine disposal on the environment.

The Egyptian Environmental Affairs Agency is not regulating and monitoring the current water management situation in Sharm sufficiently. In the longer term, this situation is likely to lead to further water shortages and price rises as well as environmental degradation which would impact on the tourism industry (El Fadel and Alameddine, 2005). Measures including stronger enforcement of relevant laws and regulations; evaluating the impact of establishing (semi-) centralized desalination and wastewater treatment plants; and considering alternative management methods for brine and sludge can help in achieving a more sustainable water resources management.

- *Which potable and non-potable water supply options are relevant to tourism-dominated arid coastal regions, with the example of Sharm El Sheikh? What are the costs associated with the selected water supply options?*

Both reverse osmosis (RO) desalination and long-distance piping options are meant to provide water of potable quality. Unit production costs of RO plants in Egypt are comparable to plants of similar capacity (above 600 m^3/d) in the IDA desalination report (2006). The observation that RO desalination costs in Egypt are higher than international trends (Hafez and El Manharawy, 2002) is not valid as plant capacities were not taken into consideration (Egyptian RO plants being generally smaller than 5,000 m^3/d). However, it should be noted that current standard industry unit production costs are much lower (about 0.5 to 1 US$/m^3; see (GWI, 2006)) than quoted figures in this research. Most of the quoted desalination plants in this research are quite old (they were built in the 1990s), they are small in size, and recently specific energy consumption has been remarkably reduced by the use of isobaric pressure exchange - energy recovery system which could also be applicable for small RO systems.

Actual O&M costs in Egypt can constitute more than 60% of the unit production cost. It is important to consider the O&M costs relative to capacity/output of the plant when estimating costs of future desalination plants.

Basic empirical cost equations were proposed to estimate RO costs and costs of long-distance piping in Egypt. The boundary of applicability for the RO cost equations is a plant capacity between 250 m^3/d to 50,000 m^3/d, and for the long distance piping equations the maximum capacity considered was 65,000 m^3/d, with a maximum length of 450 km.

RO desalination can be cheaper than long distance piping depending on capacity (of RO plant or pipeline) and length of pipeline. For instance, unit capital cost of RO desalination is cheaper than a pipeline of 140 km or longer at any capacity. For unit production cost, RO desalination is cheaper than a pipeline of length 300 km and a capacity of 2000 m^3/d. Unit production cost of RO desalination can be also competitive at a capacity of 500 m^3/d for pipelines longer than 350 km.

However, RO desalination can have some negative environmental impacts which have to be examined before considering the construction of new plants. More sustainable water supply options such as reuse of treated wastewater and water demand management should also be considered for the development of South Sinai and similar regions.

- *How can energy costs impact the costs of RO desalinated water? Can renewable sources of energy be an option?*

Desalination can be performed using thermal or membrane processes (reverse osmosis). RO desalination is the prevailing desalination technology in Egypt using conventionally generated electricity (from oil or natural gas) (Abou Rayan *et al.*, 2001; Elarabawy *et al.*, 2000). The low price of natural gas in Egypt discourages the use of renewable energy. However, as Egypt has joined the Kyoto agreement, efforts are being made to reduce the per capita CO_2 emissions. In this research, the focus was on solar thermal and solar photovoltaic (PV) systems to generate electricity for RO plants. An empirical equation is proposed to calculate the unit production cost of RO desalination depending on plant capacity, cost of energy and specific electricity demand.

Doubling the electricity prices would increase the RO unit production costs within a range of 12–40%, irrespective of plant capacity. The larger RO plants in our comparison of 21 plants in the Middle East region did not show any economies of scale in terms of annual electricity costs; the plants' energy consumption was instead mainly affected by the energy efficiency of the plant, intake water salinity, age of the plant (the older the plant, the higher the specific energy consumption) and whether there is an energy recovery system or not.

The most common combination of solar energy and desalination in Egypt is PV–RO. However, its usage is still limited to small/pilot-scale applications in remote areas. Most of the other Arab Gulf countries depend on thermal desalination, which can be coupled directly with solar thermal energy (Abd El Rassoul, 1998). When comparing the costs of solar thermal to PV, PV systems are more costly than solar thermal systems at the moment.

In fact, solar thermal costs are competitive with local energy prices and can be readily utilized. However, PV can be applied for small to medium projects (RO plant capacity of up to 15,000m^3/d) and is more popular in Egypt for that reason. It can also be readily connected to an RO desalination plant easing its implementation, while solar thermal requires a transitional phase (steam turbines to convert heat energy into electricity).

It appears that due to the reduced prices for solar energy generation modules, there is a realistic potential to make RO plants with solar energy use a viable option for Egypt and other countries in North Africa and the Middle East in the future. If the government of Egypt aims to tackle the increasing water shortage by shifting towards centralized desalination plants, this would favour the adoption of solar thermal as the most economic renewable energy source.

- *What are the factors influencing water demand in a hotel?*

Water consumption of hotels fluctuates throughout the year. From the case study it was observed that specific water consumption depends on the occupancy rate. The higher the occupancy rate, the lower the specific water consumption per room. The model calculates current and future water demands considering different scenarios of growth and water demand management.

- *What is the optimum capacity of an RO desalination plant that ensures continuous supply and minimizes water production costs?*

A model was developed using Excel macros to perform dynamic optimization with the objective function to maximize present value of total benefits over the lifetime of an RO desalination plant. The aim of the dynamic optimization was to solve for capacity expansion.

Unit production cost of RO desalination plants varies according to the degree of operation of the plant. When the plant is fully operational, unit production cost includes amortized cost on capital investment, in addition to costs of energy, labour, chemicals, and maintenance including membrane replacement. Even when the plant is partially operational, there are costs which are still incurred: amortized cost on capital, and labour and administration. This fact has to be taken into consideration when calculating costs of RO desalination and when deciding on plant capacity and whether peak demand should be satisfied in order to maximize net benefit.

With an RO plant designed to satisfy only the contracted-for water supply, the water company would be missing out on potential benefits that could have been obtained selling water in periods of high demand. On the other hand, sizing the RO plant to produce water to cover for peak demand means incurring additional costs as well as having the plant partially idle during periods of average or low demand.

Pricing excess water over contracted-for supply at higher rates than usual may retain an incentive to promote water conservation (Meij *et al.*, 2005) as it allows the water company to efficiently control consumption and satisfy their agreements with contracted-users. However, price elasticity has to be taken into consideration to determine impact of price change on water demand.

The model can be used to test different scenarios to capture time-variant tourism demand and price uncertainties on investment decisions.

- *Given the variability in water demand by a hotel, what would be the optimal contractual agreement between a hotel and a water company for both potable and non-potable water supply?*

Hotels buy their water from a water company based on a contract which stipulates a minimum daily amount of water (contracted-for water supply). Water in excess of that amount can be supplied by the water company if available. As the water company usually has contracts for the total water produced from its plant, there is rarely any excess water and hotels will then only have the option of buying from water trucks at significantly higher prices.

Hotels have to carefully analyze their water requirements in order to determine which percentage of the peak water demand should be used in the contract ("contracted-for water supply"). If the percentage chosen is high, hotels may avoid water shortages; if chosen lower, they will save costs. For potable water supply, hotels typically opt to contract for 75-80% of the peak daily domestic water demand based on experience. This research presented a methodology to optimize the contracted-for potable and non-potable water supply for hotels. It analyzed what the contracted-for water supply of a given hotel should be, based on the hotel's characteristics on one hand and the unit prices of different types of water on the other hand.

The results showed that contracted-for potable water supply is a function of several parameters including occupancy rate, size of green area and type of water used for irrigation. Hotels with expected high occupancy rates (74% and above) can contract for more than 80%. If they contract for less, this would imply higher costs in terms of having to buy excess water at a higher price in periods of high demand. On the other hand, hotels with expected lower occupancy rates (60% and less) can contract for less than 70% of the peak daily domestic water demand. The yearly savings for hotels can range from about 5-15% of the total cost of water supply.

For irrigation water supply (and similar to potable water supply), hotels buy treated wastewater from a water company based on a contract which stipulates a minimum daily amount of water (contracted-for irrigation water supply). Water in excess of that amount can be supplied by the water company if available or complemented from expensive desalinated water.

Contracting for 100% of peak irrigation demand is the highest total cost option for a hotel. Contracting for a portion of the peak irrigation demand (calculated from the optimization model) and complementing the rest from desalinated water is a cheaper option. For that reason, considering excess irrigation demand (not covered by non-potable water supply contract) while agreeing on the potable water supply contract, can minimize the risk of water shortage for irrigation.

A better option still is to satisfy excess irrigation demand from the contracted-with wastewater treatment company, if available, or from another wastewater treatment company at a higher cost (but at a cost cheaper than desalinated water). In some cases however, like in Sharm, there is only one centralized wastewater treatment plant available and the hotel has no choice but to partially use desalinated water for irrigation.

- *What is the size of a green area of a hotel that can minimize dependence on external sources of water supply and be sustained by available grey water from the hotel? How can irrigation water demand be minimized for a hotel?*

Having an on-site wastewater treatment plant presents further savings to a hotel amounting to 14% for the two example green areas in this research (Area 1 - without golf course: 57,000 m^2 and Area 2 – with golf course: 420,000 m^2). By having a golf course (Area 2), the total cost of water for a hotel at least triples.

Maximum available grey water produced by a hotel is approximately 0.7 m^3 per occupied room per day and increases to 1 m^3 per occupied room per day if there is staff housing on hotel premises. If a hotel is just to use treated wastewater on-site (without buying any excess water), it can afford about 70 (100) m^2 of green area per occupied hotel room at a conventional irrigation water requirement of 0.01 m^3/m^2/day. If drought resistant vegetation is planted and the most efficient irrigation methods are adopted, this area may double. This may not be sufficient to allow for the water requirements of a golf course, but in the case without a golf course a cost saving of 26% can be achieved.

All wastewater should be treated and re-used for irrigation, such that expensive desalinated water (whose production can cause several negative environmental impacts) will not be used for this low-value usage. The trend to construct conventional green golf courses should be halted as it requires huge amounts of blue water (desalinated water), since treated wastewater will never be sufficient.

The least costly option for a hotel is to rely solely on treated wastewater on-site. This can be achieved by: increasing the efficiency of the irrigation system, reducing the area of high-water consuming plantation (e.g. turf grass) and/or shifting to drought resistant plants including less water-consuming grass varieties. It should be noted that demand management of potable water may impact negatively on the availability of grey water for green areas. This has to be taken into consideration when designing the irrigation area in terms of area and type of plantation.

- *What are the selection criteria for the brine disposal method from an RO plant taken into consideration cost and environmental impact?*

Desalination technology has contributed to alleviating water shortages in many regions, but there are nevertheless associated negative environmental impacts, including high energy consumption and the potential surface and groundwater contamination caused by reject brine disposal. Chemicals used in pre- and post-treatment within the RO desalination process enrich reject brine with ions. The common notion is to dispose of brine as a waste product. However, brine could also be seen as a potential source of useful salts and chemicals.

The cost of reject brine disposal depends on the characteristics of the brine, level of treatment before disposal, means of disposal, volume of brine to be disposed of and nature of the disposal environment. The research estimated the direct and indirect (including environmental) costs of the disposal of reject brine through three different disposal methods. The direct cost of disposing reject brine through discharge into the sea or into a well is approximately 0.05-0.06 US$/m^3 of product water. The direct cost of disposing reject brine by means of evaporation ponds is much larger and is estimated at 0.56 US$/m^3. However, the latter technology is the only method that allows for resource recovery and combined use with fish farming may create significant benefits.

The indirect costs associated with these three methods are uncertain and largely depend on whether environmental damages will or will not occur. If brine is discharged into wells, the issue is whether aquifer contamination can be excluded or not. This depends on the local geology and the non-connectivity of the disposal well with the source water aquifer. Since the latter is often difficult to establish, the possible indirect cost may be high. In case of direct disposal of brine into the sea, the damage this may cause on sensitive marine ecosystems is potentially high and thus the potential indirect costs. The indirect cost of evaporation ponds, in contrast, is considered to be low. The following are recommendations to reduce the potential negative environmental impact of brine disposal:

1. Whereas the direct economic costs of brine disposal vary according to method and location (e.g. whether the plant is inland or coastal), the potential environmental costs are likely to be high for the methods whose direct costs are lowest. The selection of the most appropriate method thus depends on a correct estimation of the environmental costs associated with the three methods.
2. If only direct costs are taken into account, and it is not feasible to use evaporation ponds productively (fish farming and/or chemical recovery), and if the desalination plant is located near the sea, than surface discharge into the sea is the cheapest option. If, however, the indirect costs are adequately accounted for, evaporation ponds may be a more suitable option, especially where the value of land is (expected to remain) low.
3. Consistent with the above environmental considerations is the philosophy of (near) zero discharge and resource recovery. The only disposal method that would allow for resource reuse is brine disposal through evaporation ponds or brine concentrators/crystallizers.

Implications

The principal contribution of this research is a technical-economic evaluation model, which integrates most aspects of assessing the appropriateness of integrated water resources management projects. The water resources management set-up in tourism-dominated arid coastal regions was identified including potable and non-potable water supply options and a methodology for their evaluation formulated. This research represents a first attempt to provide a tool to systematically include different decision parameters to investigate the impact of variations of these parameters on the financial and environmental sustainability of a certain project taking into account its direct as well as indirect costs, which include the costs associated with possible environmental impacts.

Limitations

The developed technical-economic tool is restricted to integrated water resources management and focuses on tourism-dominated arid coastal regions. Water supply options that were considered were limited to reverse osmosis desalination and long-distance piping. Rainfall is not considered as a water supply option as the research is focused on arid regions with rainfall as low as 20 mm/yr.

It is necessary to bear in mind that the tool was developed using data from one city (Sharm El Sheikh) representing an example of tourism-dominated arid coastal regions. Though a similar set-up exists in other tourism-dominated arid coastal regions, local differences should be taken into consideration when using the tool; for instance, specific water consumption, evaporation rates, type of wastewater treatment technology, associated costs, etc.

For the validation of the model, data from existing projects from only one water company was obtained. Additional water companies declined to contribute to the research due to confidentiality concerns. In addition, public information on the case study area is very limited and very little is published. This fact contributed to the limitation of available data used to develop the model. Many data were obtained through personal communication and interviews with key personnel.

During the development process of the technical-economic modeling tool, some assumptions were used in many instances. Although in all cases these assumptions were justified, the user of the tool has to be aware of the assumptions taken and whether it is applicable to the particular case study being investigated.

Recommendations for further research

There are a number of opportunities for further research. Some opportunities of research follow from the limitations of the work presented. The most obvious is to adapt the presented technical-economic modeling tool to other regions, probably with variations in climate and water resources options from the case study area.

In the author's opinion, the most tempting and important future task would be to use the modeling tool to assess large numbers of projects by different water companies, completed and under implementation, in order to further validate the model and enhance the assumptions used.

The ongoing improvement of the technical economic modeling tool based on feedback from users is desirable, not only to assess the validity of some assumptions made, but also to reduce the uncertainties due to high variations in cost (e.g. energy prices) and in water demand.

Energy use is an issue that needs to be integrated into the discourse on sustainable tourism (Gossling, 2000). Investigating the synergies between solar energy, desalination, sustainable tourism and food production can further contribute to the overall sustainability of IWRM projects.

References

Abd Al Latif, M. (2008) Personal communication, Al Ta'meer desalination plant, Sharm El Sheikh, Egypt. Al Ta'meer desalination plant, Sharm El Sheikh, Egypt

Abd El Rassoul, R. (1998) Potential for economic solar desalination in the Middle East. *Renewable Energy*, **14**(1-4), 345-349.

Abou Rayan, M., Djebedjian, B., and Khaled, I. (2001) Water supply and demand and a desalination option for Sinai, Egypt. *Desalination*, **136**, 73-82.

Abou Rayan, M., Djebedjian, B., and Khaled, I. (2003) *Database establishment for the evaluation of the effectiveness and performance of desalination equipment in the Arab Republic of Egypt. Report*. Mansoura University, Mansura, Egypt.

Abou Rayan, M., Djebedjian, B., and Khaled, I. (2003b) Desalination option within water demand management and supply for Red Sea coast in Egypt. *Seventh international water technology conference*, Cairo, Egypt.

Abou Zeid, A. E. M. (2006) Personal communication. Eshpetco Petroleum Company. Eshpetco Petroleum company.

Abu Arabi, M. (2004) Promotion of solar desalination in the MENA region. *Middle East and North Africa renewable energy conference*, Sana'a, Yemen, 21-22 April.

Ahmed, M., Arakel, A., Hoey, D., and Thumarukudy, M. (2003) Feasibility of salt production from inland RO desalination plant reject brine: a case study. *Desalination*, **158**, 109-117.

Ahmed, M., Shayya, W., and Hoey, D. (2000) Use of evaporation ponds for brine disposal in desalination plants. *Desalination*, **130**, 155-168.

Ahmed, M., Shayya, W., Hoey, D., and Al Handaly, J. (2001) Brine disposal from reverse osmosis desalination plants in Oman and the United Arab Emirates. *Desalination*, **133**, 135-147.

Ahmed, S., Tewfik, S., and Talaat, H. (2002) Development and verification of a decision support system for the selection of optimum water reuse schemes *Desalination*, **12**, 339-352.

Al Agha, M., and Mortaja, R. (2005) Desalination in the Gaza strip: drinking water supply and environmental impact. *Desalination*, **175**, 157-171.

Al Suleimani, Z., and Nair V., R. (2000) Desalination by solar-powered reverse osmosis in remote area of the Sultanate of Oman. *Applied Energy*, **65**(1), 367-380.

Al Zubari, W. (2003) *Alternative water policies for the Gulf cooperation council countries*, A.S. Alsharhan and W.W. Wood, Elsevier Science, Amsterdam, The Netherlands,

Alvarea, D., and Lourenco, J. (2005) *Life cycle modeling for tourism areas, Thesis*. University of Minho, Portugal. [Online] http://repositorium.sdum.uminho.pt/bitstream/1822/4982/1/Lourenco_CI_3_2005.pdf

Anderson, J., and Iyaduri, R. (2003) Integrated urban water planning: big picture planning is good for the wallet and the environment. *Water science and technology*, **47**(7), 19-23.

Avlonitis, S. A., Poulios, I., Vlachakis, N., Tsitmidelis, S., Kououmbas, K., Avlonitis, D., and Pavlou, M. (2002) Water resources management for the prefecture of Dodekansia of Greece. *Desalination*, **152**, 41-50.

Bahri, A., Basset, C., Oueslati, F., and Brissaud, F. (2001) Reuse of reclaimed wastewater for golf course irrigation in Tunisia. *Water Sci. & Tech.,* 43(10), 117-124.

Baker, R. (2006) *Membrane technology and applications*, John Wiley & Sons, Ltd, England, 2nd edition.

Barbagallo, S., Luigi, G., Consoli, S., and Somma, F. (2003) Wastewater quality improvement through storage: a case study in Sicily. *Water Sci. & Tech.*, **47**(7), 169-176.

Brimberg, J., Oron, G., and Mehrez, A. (1993) A model for the development of marginal water sources in Arid Zones: the case of the Negev Desert, Israel. *Water Resources Research*, **29**(9), 3059-3067.

CEEI (2001) *In depth analysis of trade and industrial barriers in environmental energy sectors-final report*. Cairo.

Chaibi, M. T. (2000) An overview of solar desalination for domestic and agriculture water needs in remote arid areas. *Desalination*, **127**, 119-133.

Chartzoulakis, K., Paranychianakis, N., and Angelakis, A. (2001) Water resources management in the island of Crete, Greece, with emphasis on the agricultural use. *Water Policy*, **3**, 193-205.

Choi, H. R., Kim, W., and Youn An, S. (1997) Recurrent and decomposed neural network-based hotel occupancy forecasting. *New Review of Applied Expert Systems*, **3**, 121-136.

CIA (2005) *Egypt people: World Fact book*. [Online] http://www.theodora.com/wfbcurrent/egypt/egypt_people.html

Crites, R., Reed, S., and Bastian, R. (2000) *Land treatment systems for municipal and industrial wastes*, McGraw-Hill,

Dakkah, M., Hirata, A., Muhida, R., and Kawasaki, Z. (2003) Operation strategy of residential centralized photovoltaic system in remote areas. *Renewable energy*, **28**, 997-1012.

Dandy, G. C., McBean, E. A., and Hutchinson, B. G. (1984) A model for constrained optimum water pricing and capacity expansion. *Water Resources Research*, **20**(5), 511-520.

Delleur, J. (2007) *The handbook of groundwater engineering*, CRC Press, United States, 2nd Edition.

Deng, S., and Burnett, J. (2000) Audit of water consumption in hotels in Hong Kong. *Building Serv. Eng. Res. Technol.*, **21**(3), 209-211.

Directory, I. (2005) *News from print, Water and Wastewater news*. [Online] http://wwn-online.com/articles/51176/

Downs, T., Mazari, M., Mora, R., and Suffet, I. (2000) Sustainability of least cost projects for meeting Mexico's city future water demand. *Water resources research*, **36**(8), 2321-2339.

Dube, E., and Van der Zaag, P. (2003) Analyzing water use patterns for demand management: the case of the city of Masvingo, Zimbabwe. *Physics and Chemistry of the Earth*, **28**, 805-815.

Dunlop, J. P., Farhi, B. N., Post, H. N., Szaro, J. S., and Ventre, G. G. (2001) Reducing the costs of grid connected photovoltaic systems. *Solar forum 2001: Solar energy, the power to choose* Washington, D.C., 21-25 April.

Durham, B., Yoxtheimer, D., Alloway, C., and Diaz, C. (2003) Innovative water resources solutions for islands. *Desalination*, **156**, 155-161.

EDGAR (2007) *Over 30 years of dependable water production and distribution, Consolidated Water Co. LTD, The Cayman Islands*. [Online] http://ir.cwco.com/

EEDRB (2005) *CO2 emissions fro Arab Republic of Egypt, International Atomic Energy Agency*. [Online] http://www.iaea.org/inisnkm/nkm/aws/eedrb/data/EG-enemc.html

Einav, R., Harussi, K., and Perry, D. (2002) The footprint of the desalination process on the environment. *Desalination*, **152**, 141-154.

Einav, R., and Lokiec, F. (2003) Environmental aspects of a desalination plant in Ashkelon. *Desalination*, **156**, 79-85.

El Fadel, M., and Alameddine, I. (2005) Desalination in arid regions: Merits and concerns. *Journal of water supply: research and technology - Aqua*, **54**, 449-461.

El Nokrashy, H. (2005) Renewable mix for Egypt. *Cairo 9th International Conference on Energy and Environment*, Cairo, Egypt.

El Nokrashy, H. (2006) Personal communication. Nokrashy Engineering GMBH. Germany.

Elarabawy, M., Attia, B., and Tosswell, P. (2000) Integrated water resources management for Egypt. *Journal of water supply: research and technology - Aqua*, **49.3**, 111-125.

ESIS (2008) *Egypt state information service*. [Online] http://www.sis.gov.eg.

Fiorenza, G., Sharma, V. K., and Braccio, G. (2003) Techno-economic evaluation of a solar powered water desalination plant. *Energy conservation and management*, **44**, 2217-2240.

Froukh, M. (2001) Decision support system for domestic water demand forecasting and management. *Water resources management*, **15**, 155-161.

Gasco, G. (2004) Influence of state support on water desalination in Spain. *Desalination*, **165**, 111-122.

Girgis, Z. (2008) Personal Communication, CEO Ridgewood Egypt for infrastructure projects (mainly water and power).

Glater, J., and Cohen, Y. (2003) *Brine disposal from land based membrane desalination plants, a critical assessment*. Metropolitan water district of Southern California.

Glueckstern, P., and Priel, M. (1996) Optimized brackish water desalination plants with minimized impact on the environment. *Desalination*, **108**, 19-26.

Gonzalez, E., Rodriguez, J., Cordero, T., Koussis, A., and Rodriguez, J. (2005) Cost of reclaimed municipal wastewater for applications in seasonally stressed semi-arid regions. *Journal of water supply: research and technology - Aqua*, **54.6**, 355-369.

Gopalakrishnan, C., and Cox, L. (2003) Water consumption by the visitor industry: the case of Hawaii. *Water resources development*, **19**(1), 29-35.

Gordan, D. (2001) *Incorporating environmental costs into an economic analysis of water supply planning: a case study of Israel*, Simon Fraser University, British Columbia, Canada.

Gossling, S. (2000) Sustainable tourism development in developing countries: some aspects of energy use. *Journal of sustainable tourism*, **8**(5), 410-425.

GWI (2006) *The 2006 IDA World wide Desalination Plant Inventory Report Number 19*.

Hafez, A., and El Manharawy, S. (2002) Economics of seawater RO desalination in the Red Sea region, Egypt. Part 1. A case study. *Desalination*, **153**, 335-347.

Hamoda, M. (2004) Water strategies and potential of water resources in the South Mediterranean countries. *Desalination*, **165**, 31-41.

Hinomoto, H. (1972) Dynamic programming of capacity expansion of municipal water treatment systems. *Water Resources Research*, **8**(5), 1178-1187.

Hoffmann, W. (2006) PV solar electricity industry market growth and perspective. *Solar energy materials and solar cells*, **90**, 3285-3311.

Kally, E. (1993) *Water and peace: water resources and the Arab-Israeli peace process*. Greenwood, Oxford, UK.

Karagiannis, C., Banat, F., Fath, H., and Mathioulakis, E. (2007) Technical and economic analysis of Autonomous desalination systems: case studies for eleven ADIRA installations in Middle East and North African countries. *MEDA Water International conference*, Tunis, Tunisia, 21-24 May.

Khaled, I. (2008) Personal communication, Director of Sinai Development Authority, South Sinai, Egypt. Sinai Development Authority, South Sinai, Egypt.

Khalil, E. (2004) Water strategies and technological development in Egyptian coastal areas. *Desalination*, **165**, 23-30.

Khouzam, R. (2003) Economic aspects of wastewater reuse, case study: the Arab Nation. *Economic research forum for Arab nations, Turkey and Iran*, Cairo-Egypt.

Kimes, J. (1995) The regulation of concentrate disposal in Florida. *Desalination*, **102**, 87-92.

Lamei, A., van der Zaag, P. and von Münch, E. (2009a) Water resources management to satisfy high water demand in the arid Sharm El Sheikh on the Red Sea, *Desalination & Water treatment*, **1**, 299-306.

Lamei, A. van der Zaag, P. and von Münch, E. (2008a) Basic cost equations to estimate unit production costs for RO desalination and long-distance piping to supply water to tourism-dominated arid coastal regions of Egypt. *Desalination*, **225,** 1-12.

Lamei, A., van der Zaag, P. and von Münch, E. (2008b) Impact of solar energy cost on water production cost of seawater desalination plants in Egypt. *Energy policy*, **36**, 1748-1756.

Lamei, A., von Münch, E., Imam, E., and van der Zaag, P. (2006) A model for calculation of water demand by the tourism industry. *Integrated Water Resources Management Conference and Challenges of the Sustainable Development (Moroccan Committee of the International Association of Hydrogeologists)*, Marrakech, Morocco, 23-25 May.

Lamei, A., von Münch, E., van der Zaag, P. and Imam, E. (2009b) Optimum contracted-for potable water supply for hotels in arid coastal regions. *Water Sci. & Tech,* **59**(8), 1541-1550.

Lamei, A., Tilmant, A., van der Zaag, P. and Imam, E. (2009c) Dynamic programming of capacity expansion of reverse osmosis desalination plant Case study: Sharm El Sheikh, Egypt. *Water Sci. & Tech.: Water Supply,* **9**(3), 233-246.

Lamei, A., Von Münch, E., Van der Zaag, P., and Emad, I. (submitted) Integrating wastewater reuse in water resources management and optimization of contracted-for wastewater supply for hotels in arid coastal regions: the case study of Sharm El Sheikh, Egypt. *Water Sci. & Tec.*

Lamei, A., Von Münch, E., and Van der Zaag, P. (2009d) Environmental impact and economic costs of brine disposal methods from RO desalination plants in arid coastal regions. Accepted paper prepared for the *IDA Congress, Dubai, UAE,* 7-12 November.

Latorre, m. (2005) Environmental impact of brine disposal on Posidonia seagrasses. *Desalination*, **182**, 517-524.

Lazarova, V., Levine, B., Sack, J., Cirelli, G., and Jeffrey, P. (2001) Role of water reuse for enhancing integrated water management in Europe and Mediterranean countries. *Water Sci. & Tech.*, **43**(10), 25-33.

Loetscher, T. (2000) *WaterGuide, Advanced wastewater management centre. The University of Queensland and Queensland government, Australia*

Loucks, D., and Van Beek, E. (2005) *Water resources systems planning and management: an introduction to methods, models and applications*, UNESCO, Paris,

Mahmoud, M., Fahmy, H., and Laborice, J. (2002) Multi-criteria sitting and sizing of desalination plants with GIS. *Journal of water resources planning and management*, **128**(2), 113-120.

Meade, B. (1998) *Improving water use efficiency in Jamaican hotels and resorts through the implementation of environmental management systems. Hagler Baily Services, Inc. .* [Online] http://www.ucowr.siu.edu/updates/pdf/v115_A6.pdf

Meij, S., Stumphius, J. C., and Ruiters, C. J. M. (2005) Market-driven pricing structures for drinking water. *Water Sci. & Tech.: Water Supply*, **5**(6), 225-233.

Mohsen, N. (2007) Personal communication. Fantazia hotel, Sharm El Sheikh, Egypt.

MSEA (2006) *Law 4 for the protection of the environment.* [Online] http://www.eeaa.gov.eg/English/main/about.asp

MWRI (2005) *National water resources plan for Egypt 2017, Ministry of water resources and irrigation*. Cairo.

OECD (1994) *Project and policy appraisal: Integrating economics and environment. Paris.*

Oron, G. (1996) Management modeling of Integrative wastewater treatment and reuse systems. *Wat. Sci Tech.*, **33**(10-11), 95-105.

Paraskevas, P. A., Giokas, D. L., and Lekkas, T. D. (2004) Non-conventional water resources in coastal urban areas, the case of Greece. *Water science and technology*, **46**(8), 177-186.

Pearce, D., and Markandya, A. (1989) *Marginal opportunity costs as a planning concept in natural resources management. Environmental management and economic development*, Baltimore. John Hopkins University Press,

Ramjeawon, T. (1994) Water resources management on the small island of Mauritius. *Water resources development*, **10**(2), 143-155.

Ravizky, A., and Nadav, N. (2007) Salt production by the evaporation of SWRO brine in Eilat: a success story. *Desalination*, **205**, 374-379.

Saad, R. (2005) *Mixed signals, Al Ahram Weekly.* [Online] http://weekly.ahram.org.eg/2005/753/fo5.htm

Sadi, A. (2004) Seawater desalination share among water and market policy changes in Algeria. *Desalination*, **165**, 99-104.

Saleh, M. (2008) Personal communication, Managing Director, Global Energy, Cairo, Egypt.

Salgot, M., and Tapias, J. (2004) Non-conventional water resources in coastal areas: a review on the use of reclaimed water. *Geologica Acta*, **2**(2), 121-133.

Sanden, B., and Azar, C. (2005) Near-term technology policies for long-term climate targets-economy wide versus technology specific approaches. *Energy policy*, **33**, 1557-1576.

Sauer, J. (2005) Economies of scale and firm size optimization in rural water supply. *Water Resources Research*, **41**(W11418, doi:10.1029/2005WR004127).

Savenije, H., and van der Zaag, P. (2008) Integrated water resources management: concepts and issues. *Physics and chemistry of the earth*, **33**, 290-297.

Schachtschneider, K. (2000) Water demand management study of Namibian tourist facilities. *1st WARFSA/ WaterNet Symposium: sustainable use of water resources*, Maputo, 1-2 Nov.

Semiat, R. (2000) Desalination: Present and Future. *Water International*, **25**(1), 54-65.

Seppala, O. T., and Katko, T. S. (2003) Appropriate pricing and cost recovery in water services. *Journal of water supply: research and technology - Aqua*, **52.3**, 225-236.

Shahatto, S. (2003) *Environmental impact of development in coastal areas: desalination, PhD Thesis*, University of Tubingen, Germany.

Shelef, G., and Azov, Y. (1996) The coming era of intensive wastewater reuse in the Mediterranean region. *Water Sci. & Tech.*, **33**(10-11), 115-125.

Sipala, S., Mancini, G., and Vagliasindi, F. G. A. (2003) Development of a web-based tool for the calculation of costs of different wastewater treatment and reuse scenarios. *Water Sci. & Tech.: water supply*, **3**(4), 89-96.

Snow, J. (2001) *Water conservation on golf courses, United States golf association, International turf producers association.* [Online] http://www.usga.org/turf/articles/environment/water/water_conservation.html

Sommariva, C. (2004) *Desalination management and economics*, Faversham House Group, UK,

Sonune, A., and Ghate, R. (2004) Development in wastewater treatment technologies. *Desalination*, **167**, 55-63.

Strategan (2004) *Metropolitan desalination proposal, section 46 review, Australia*. Australia.

Svensson, M. (2005) *Desalination and the environment: options and considerations for brine disposal in inland and coastal locations*. Department of biometry and engineering, SLU.

Tanik, A., Ekdal, F., Germirli, B., and Orhon, D. (2005) Recent practices on wastewater reuse in Turkey. *Water Sci. & Tech.,* **51**(11), 141-149.

Tchobanoglous, G., Burton, F., and Stensel, H. (2003) *Wastewater engineering treatment and reuse*, Metcalf and Eddy, Inc., 4th Edition.

Thomas, J., and Durham, B. (2003) Integrated water resources management: looking at the whole picture. *Desalination*, **156**, 21-28.

Trung, D., and Kumar, S. (2005) Resource use and waste management in Vietnam hotel industry *Journal of cleaner production*, **13**, 109-116.

UNEP (2000) *Source book of alternatives for freshwater augmentation in some countries in Asia.* [Online] http://www.unep.or.jp/ietc/publications/techpublications/techpub-8e/desalination.asp

UNEP/PERSGA (1997) *Assessment of land-based sources and activities affecting the marine environment in the Red sea and Gulf of Aden, UNEP Regional Seas Report and Studies, No. 166, United Nations Environmental program* [Online] http://www.unep.ch/regionalseas/main/persga/redthreat.html

Unknown (1998) *Energy Information Administration, United States. Egypt.* [Online] http://www.converger.com/eiacab/egypt.htm

Unknown (2008) *Restore and monitor water quality, City of Tacoma, Washington State, USA.* [Online] http://www.cityoftacoma.org/Page.aspx?hid=939

US (2006) *Energy efficiency and renewable energy.* US Department of Energy. [Online] http://www.eere.energy.gov/solar/cfm/faqs/third_level.cfm/name=Photovoltaics/cat=The%20Basics

USCongress (1988) *using desalination technologies for water treatment, OTA-BP-O-46, Washington, DC.* US government printing office, Washington, DC.

USDI (2002) *Zero discharge waste brine management for desalination plants.* Desalination research and development program, Report no. 89, El Paso, Texas. http://www.usbr.gov/pmts/water/publications/reportpdfs/report089.pdf, El Paso, Texas.

USEPA (1992) *Economic Analysis of pollution prevention project, EPA/600/R-92/088, http://www.epa.state.oh.us/opp/tanbook/fppgch6.txt*.

Van der Merw, B. (1999) *IUCN water demand management country study-Namibia.* . Directorate Resource Management DWA, MAWRD and city engineer (water Services) Windhoek

Vedavyasan, C. V. (2001) Combating brine disposal under various scenarios. *Desalination*, **139**, 419-421.

Voivontas, D., Arampatzis, G., Maoli, E., Karavitis, C., and Assimacopoulous, D. (2003) Water supply modeling towards sustainable environmental management in small islands: the case of Paros, Greece. *Desalination*, **156**, 127-135.

WC (2004) *Kimberley pipeline project, Sustainability Review, Water Corporation. 61/15328/46718*, Perth, Australia.

WDD (2007) *Water works. Desalination plants. Water Development Department. Cyprus.* [Online] http://www.cyprus.gov.cy/moa/wdd/WDD.nsf

WTO (2007) *Yearbook of tourism statistics Data 2001-2005*, World Tourism Organization,

XIn, H. (2006) *Egypt expects quick recovery of tourism industry following April terror attacks*. People's daily online. [Online] http://www.english.peopledaily.com

Xu, P. (2002) *Technical and economic modeling of water resources management integrating wastewater reuse, PhD Thesis*. Ecole Nationale du Genie Rural des Eaux et des Forets, Montpellier, France, France.

Xu, P., Brissaud, F., and Salgot, M. (2003a) Facing water shortage in a Mediterranean tourist area : seawater desalination or water reuse? *Water Sci. & Tech.: Water Supply*, **3**(3), 63-70.

Xu, P., Valette, F., Brissaud, F., Fazio, A., and Lazaova, V. (2003b) Technical-Economic modeling of integrated water management: wastewater reuse in a French island. *Water Sci. & Tech.*, **43**(10), 67-74.

Young, R. (2005) *Determining the economic value of water, concepts and methods*, Resources for the future, Washington, D.C. USA,

Younos, T. (2005) Environmental issues of desalination. *Journal of contemporary water research and education*, **132**, 11-18.

Youssef, R. (2006) *Desalination technology roadmap 2030.* Centre for future studies, the Cabinet of Information and Decision Support Centre, Cairo.

Zhou, Y., and Tol, R. (2005) Evaluating the costs of desalination and water transport. *Water resources research*, **41**, W03003.

Summary

Water scarcity aggravates in coastal zones which are characterized by high population density, intense economic activity and tourism, and consequently facing a heavy seasonally variable water demand. The usual way to tackle water demand is to attempt to satisfy it through adopting supply-side measures such as increasing the abstraction from conventional surface and ground water sources. However, due to increasing limitation in water resources, a shift is taking place towards integrated water resources management (IWRM).

IWRM adopts a holistic approach to optimize water usage. IWRM has to take into account the following four dimensions: water resources, water users, spatial and temporal scales. It critically assesses supply options, including developing alternative water resources, e.g. desalination (removal of dissolved minerals including salts) and reclaimed wastewater. It also seeks to increase the management efficiency of conventional resources and schemes. Finally, and importantly, an IWRM approach will also consider demand management options. IWRM projects should be sustainable and fulfil the public interest: maintenance of environmental quality, financial sustainability, good governance, institutional capacities and social equity. However, the development, decision making process and implementation of projects conforming to the IWRM approach are complex due to the different sectors involved, typically water, environment, energy and agriculture.

The scope of this study is to develop a technical-economic modeling tool to aid decision makers, both public and private sector investors, in the design and assessment of the financial and environmental sustainability of integrated water resources management projects for tourism-dominated arid coastal regions. Designed projects are to satisfy varying temporal and spatial demands and to find sustainable solutions, which may include a higher level of wastewater reuse. The decision maker using the technical-economic model can be a hotel, a water company, or a water management company (i.e. municipality, city, town, and governorate). The water management company is assumed to perform integrated water management including: supply of potable water, collection and treatment of wastewater, and reuse of treated wastewater for irrigation water supply

The modeling tool box is composed of five types of models depending on the specific site situations: water demand/need model, water supply model, wastewater reuse model, environmental model and economic model. The coupling of the models facilitates assessing different scenarios for integrated water resources management including different time-variant domestic and irrigation water demands, water supply options: long-distance piping and reverse osmosis desalination (RO), potable and non-potable water supply contracts between a hotel and a water company, optimum capacity and expansion for an RO desalination plants, renewable and non-renewable energy sources for RO desalination, and brine disposal options. The different scenarios are compared according to a cost benefit analysis and their potential environmental impact (where applicable). The models are developed using Excel Macros.

The modeling tool is applied to address water management issues in the case study region (Sharm El Sheikh, South-Sinai, Egypt) that depends heavily on tourism. The city suffers from water shortage and is located near a unique but sensitive marine environment of high ecological value. Water supply to Sharm is mainly from RO desalination complemented by groundwater from Al Tor which is transported by tankers or long-distance pipelines as well as from treated domestic wastewater (for landscape irrigation).

RO desalination can be cheaper than long distance piping depending on capacity (of RO plant or pipeline) and length of pipeline. For instance, unit capital cost of RO desalination is cheaper than a pipeline of 140 km or longer at any capacity. For unit production cost, RO desalination is cheaper than a pipeline of length 300 km and a capacity of 2000 m^3/d. Unit production cost of RO desalination can be also competitive at a capacity of 500 m^3/d for pipelines longer than 350 km.

However, RO desalination can have some negative environmental impacts which have to be examined before considering the construction of new plants including potential environmental impact from brine disposal and high energy consumption. The research estimated the direct and indirect (including environmental) costs of the disposal of reject brine through three different disposal methods: surface discharge, well disposal and evaporation ponds. The direct cost of disposing reject brine through discharge into the sea or into a well is approximately 0.05-0.06 US$/$m^3$ of product water. The direct cost of disposing reject brine by means of evaporation ponds is much larger and is estimated at 0.56 US$/$m^3$. However, the latter technology is the only method that allows for resource recovery and combined use with fish farming may create significant benefits. The indirect costs associated with these three methods are uncertain and largely depend on whether environmental damages will or will not occur.

Solar thermal costs for RO desalination are competitive with local energy prices and can be readily utilized. However, Photo-Voltaic cells can be applied for small to medium projects (RO plant capacity of up to 15,000m^3/d) and are more popular in Egypt for that reason. They can also be readily connected to an RO desalination plant easing its implementation, while solar thermal requires a transitional phase (steam turbines to convert heat energy into electricity). It appears that due to reducing prices for solar energy generation modules, there is a realistic potential to make RO plants with solar energy use a viable option for Egypt and other countries in North Africa and the Middle East in the future. If the government of Egypt aims to tackle the increasing water shortage by shifting towards centralized desalination plants, this would favour the adoption of solar thermal as the most economic renewable energy source.

Water consumption of hotels fluctuates throughout the year. From the case study it was observed that specific water consumption depends on the occupancy rate. The higher the occupancy rate, the lower the specific water consumption per room. The model calculates current and future water demands considering different scenarios of growth and demand management.

Hotels buy their potable and non-potable water from a water company based on a contract which stipulates a minimum daily amount of water (contracted-for water supply). Water in excess of that amount can be supplied by the water company if available. As the water company usually has contracts for the total water produced from its plant, there is rarely any excess water and hotels will then only have the option of buying from water trucks at significantly higher prices.

Contracted-for potable water supply is a function of several parameters including occupancy rate, size of green area and type of water used for irrigation. Hotels with expected high occupancy rates (74% and above) can contract for more than 80% of the peak daily domestic water demand. If they contract for less, this would imply higher costs in terms of having to buy excess water at a higher price in periods of high demand. On the other hand, hotels with expected lower occupancy rates (60% and less) can contract for less than 70% of the peak demand.

Maximum available grey water produced by a hotel is approximately 0.7 m^3 per occupied room per day and increases to 1 m^3 per occupied room per day if there is staff housing on the hotel premises. If a hotel is just to use treated wastewater on-site (without buying any excess water), it can afford about 70 (100) m^2 of green area per occupied hotel room at a conventional irrigation water requirement of 0.01 m^3/m^2/day. If drought resistant vegetation is planted and the most efficient irrigation methods are adopted, this area may double. This may not be sufficient to allow for the water requirements of a golf course, but in the case without a golf course a cost saving of 26% can be achieved. All wastewater should be treated and re-used for irrigation, such that expensive desalinated water (whose production can cause several negative environmental impacts) will not be used for this low-value usage. The trend to construct conventional green golf courses should be halted as it requires huge amounts of desalinated blue water, since treated wastewater will never be sufficient.

Finally, with an RO plant designed to satisfy only the contracted-for water supply, the water company would be missing out on potential benefits that could have been obtained selling water in periods of high demand. On the other hand, sizing the RO plant to produce water to cover for peak demand means incurring additional costs as well as having the plant partially idle during periods of average or low demand. A dynamic optimization model is developed with the objective function to maximize present value of total benefits over the lifetime of an RO desalination plant. The aim of the dynamic optimization is to solve for capacity expansion.

Samenvatting

Water schaarste wordt steeds ernstiger in kustgebieden die gekenmerkt worden door hoge bevolkingsdichtheid, intensieve economische activiteit en toerisme, met als gevolg grote seizoensgebonden schommelingen in de vraag naar water. De conventionele manier om aan de vraag naar water te voldoen is door middel van aanbodgestuurde maatregelen, zoals het vergroten van de inname uit conventionele bronnen van grond- en oppervlaktewater. Echter, vanwege de steeds nijpender watervoorraden vindt er een verschuiving plaats naar een integrale benadering van waterbeheer, namelijk Integrated Water Resources Management (IWRM).

IWRM benadert het optimaliseren van water gebruik op een holistische wijze, en kijkt naar de volgende vier dimensies: de water cyclus, de watergebruikers, en de ruimtelijke en temporele schalen. Het analyseert aanbodgestuurde oplossingen, inclusief het ontwikkelen van alternatieve waterbronnen, zoals ontzilting van zeewater en hergebruik van afvalwater. Het kijkt ook naar mogelijkheden om het beheer van conventionele waterbronnen en waterprojecten efficiënter te maken. Een integrale benadering neemt ook maatregelen die zich richten of the vraag naar water in aanmerking. Integrale waterprojecten zijn duurzaam en dienen de publieke zaak, waarbij waarden voor milieukwaliteit, goed financieel beheer, behoorlijk bestuur, institutionele capaciteit en sociale gelijkheid gewaarborgd worden. Het ontwikkelen en uitvoeren van projecten die voldoen aan deze integrale benadering is complex omdat er verschillende sectoren bij betrokken zijn – veelal water, milieu, energie en landbouw.

Deze studie wil een technisch-economisch instrument ontwikkelen dat beleidsmakers – van de publieke en private sector - kan ondersteunen in het ontwerpen en evalueren van de financiële en ecologische duurzaamheid van integrale water projecten voor aride kustgebieden. Succesvolle projecten moeten op een duurzame manier aan een in tijd en plaats sterk variërende vraag naar water kunnen voldoen – bijvoorbeeld door intensiever hergebruik van afvalwater. Het technisch-economisch instrument moet gebruikt kunnen worden door beleidsmakers van de hotel industrie, drinkwaterbedrijven en waterbeer instanties op gemeentelijk en provinciaal niveau. Waterbeheer instanties worden verondersteld verantwoordelijk te zijn voor integraal waterbeheer, waaronder drinkwatervoorziening, verzameling en zuivering van afvalwater, en hergebruik daarvan voor irrigatie.

Het technisch-economisch instrument omvat vijf verschillende modellen die de vraag naar water, de watervoorziening, het hergebruik van afvalwater, de milieu effecten en de financieel-economische consequenties kunnen simuleren. Door deze modellen te koppelen kunnen scenario's geëvalueerd worden die verschillen in tijdsafhankelijke vraag voor huishoudelijk en irrigatie water combineren met alternatieve watervoorziening opties, zoals lange afstandspijpleidingen en ontzilting van zeewater met behulp van de *reverse osmosis* (RO) techniek, contracten van water bedrijven met hotels voor de levering van drinkwater en (gezuiverd afval)water, de optimale capaciteit en uitbreiding van RO ontzilting installaties, het gebruik van conventionele en hernieuwbare energie voor deze installaties, en verschillende opties voor het verwijderen van het afvalproduct van RO (brijn). Scenario's kunnen vergeleken worden op basis van kostenbaten en milieueffect analyses. De modellen zijn ontwikkeld met behulp van Excel Macros.

Het technisch-economisch instrument is toegepast op de waterbeheer problematiek in het toeristisch belangrijke Sharm El Sheikh in Egypte. De stad heeft te lijden van water tekorten en ligt nabij een uniek maar delicaat marien ecosysteem van hoge ecologische waarde.

Drinkwater komt voornamelijk van RO ontzilting installaties, alsmede van grondwater uit Al Tor dat aangevoerd wordt met behulp van trucks en langeafstandspijpleidingen, en van gezuiverd afvalwater voor het irrigeren van groene zones.

RO ontzilting kan goedkoper zijn dan het aanvoeren van water van ver, afhankelijk van de capaciteit (van de RO installatie of pijpleiding) en afstand tot de waterbron. Zo zijn de investeringskosten van RO ontzilting lager dan van een pijleiding langer dan 140 km, onafhankelijk van de capaciteit. Als er gekeken wordt naar de productiekosten per volume water dan is het RO ontzilten van 2,000 m^3 zeewater per dag goedkoper dan het transporteren van dezelfde hoeveelheid zoetwater over een afstand van 300 km. Als de alternatieve zoetwaterbron zich op een afstand van 350 km bevindt dan is een RO ontzilting installatie met een capaciteit van 500 m^3 per dag al concurrerend.

RO ontzilting kan nochtans negatieve milieu effecten veroorzaken, zoals mogelijke ecologische schade door het lozen van brijn en door het relatief hoge energieverbruik, welke geëvalueerd dienen te worden vóórdat tot het bouwen van nieuwe installaties wordt overgegaan. Deze studie heeft een schatting gemaakt van de directe en indirecte (almede milieu) kosten van drie verschillende methoden om van het brijn af te komen; namelijk via oppervlakkige afvoer naar de zee, via afvoer naar waterputten (injectie in grondwater), en met behulp van verdampingsvijvers. De directe kosten om brijn te dumpen in de zee of in een put zijn 0.05-0.06 US$ per m^3 geproduceerd drinkwater. De directe kosten om afvalzout te verwijderen met behulp van verdampingsvijvers zijn veel groter, en worden geschat op 0.56 US$ per m^3 geproduceerd drinkwater. Een verdampingsvijver is echter de enige technologische oplossing met de mogelijkheid van grondstof terugwinning. In combinatie met visproductie kan dit commercieel interessant zijn. De indirecte kosten van deze drie methoden zijn grotendeels onzeker, en worden vooral bepaald of potentiële nadelige milieu effecten al dan niet zullen optreden.

De kosten van thermische zonne-energie voor RO ontzilting zijn concurrerend met de lokale prijzen van conventionele energie. PV (fotovoltaische) panelen die elektrische zonne-energie opwekken kunnen echter toegepast worden op kleine en middelgrote RO installaties (capaciteit tot 15,000 m^3/dag) en komen daarom veel voor in Egypte. PV panelen kunnen rechtstreeks aan een RO ontzilting installatie gekoppeld worden wat hun gebruik vergemakkelijkt; het gebruik van thermische zonne-energie voor RO vergt daarentegen een tussenstap (stoom turbines die warmte energie omzetten in elektriciteit). De dalende kosten van zonnecellen maken dat de nabije toekomst RO ontzilting met behulp van zonne-energie een reële optie is voor Egypte en andere landen in Noord Afrika en het Middenoosten. Indien de overheid van Egypte de steeds nijpender waterschaarste van zins is aan te pakken door het aanleggen van meer grootschalige ontzilting installaties, dan is thermische zonne-energie de meest economische duurzame energiebron.

Het watergebruik van hotels fluctueert gedurende het jaar. De geval studie heeft laten zien dat het specifieke watergebruik afhangt van de bezettingsgraad: hoe hoger de bezettingsgraad hoe lager het specifieke watergebruik per hotelkamer. Het model berekent de huidige en toekomstige waterbehoefte afhankelijk van verschillende groei en vraaggestuurde beheer scenario's. Hotels betrekken hun drinkwater en irrigatiewater van waterbedrijven op basis van contracten waarin een minimum dagelijkse hoeveelheid is vastgelegd (gecontracteerde watervoorziening). Het waterbedrijf kan water leveren boven deze vastgestelde hoeveelheid, maar alleen als dit beschikbaar is. Omdat een waterbedrijf gewoonlijk contracten heeft gesloten voor de totale leveringscapaciteit kan het vaak geen additioneel water leveren; dan hebben hotels geen andere keuze dan water van trucks te kopen voor een veel hogere prijs.

De hoeveelheid gecontracteerd drinkwater hangt af van verschillende variabelen, zoals de bezettingsgraad, de grootte van de groene (geïrrigeerde) zone, en welk type water voor irrigatie wordt gebruikt. Hotels die een hoge bezettingsgraad verwachten (meer dan 74%) doen er goed aan contracten af te sluiten voor tenminste 80% van de piek waterbehoefte. Als ze voor minder contracteren zullen ze hogere kosten maken om additioneel water te kopen gedurende periodes met grote vraag. Echter, hotels die een lage bezettingsgraad verwachten (minder dan 60%) kunnen contracten sluiten voor minder dan 70% van de piek behoefte.

De maximale hoeveelheid "grijs water" dat een hotel produceert is ongeveer 0.7 m^3 per bezettte kamer per dag, en stijgt tot 1.0 m^3 indien er ook huisvesting voor het personeel op het terrein aanwezig is. Indien een hotel alleen aangewezen is op het hergebruik van dit afvalwater (zonder additioneel water aan te kopen) kan het een groene zone van ongeveer 70 m^2 (100 m^2) per bezette hotelkamer onderhouden, uitgaande van een irrigatiewaterbehoefte van 0.01 m^3/m^2/dag. Indien droogte-resistente vegetatie wordt geplant en de meest efficiënte irrigatiemethodes worden gebruikt kan deze groene oppervlakte verdubbeld worden. Dit zal niet altijd voldoende zijn voor de waterbehoefte van een golfbaan, maar hotels zonder golfbaan kunnen een kostenbesparing van 26% bereiken. Al het afvalwater moet gezuiverd worden en hergebruikt voor irrigatie, zodat duur ontzilt water (de productie waarvan negatieve milieueffecten kan veroorzaken) niet aangewend wordt voor gebruiken met een lage economische waarde. De trend om conventionele groene golfbanen aan te leggen moet gestopt worden, omdat dit zeer grote hoeveelheden ontzilt water vereist daar er nooit voldoende gezuiverd afvalwater zal zijn.

Tenslotte, indien een RO installatie uitsluitend ontworpen is om de gecontracteerde hoeveelheid water te produceren zou het potentiële voordelen mislopen om water te verkopen in tijden van grote vraag. Echter, RO installaties ontworpen voor piekbehoeftes vergen grotere investeringen en liggen gedurende periodes met gewone en lage vraag gedeeltelijk stil. Een dynamisch optimalisatie model is ontwikkeld met als doelfunctie de contante waarde van alle baten gedurende de levensduur van de RO ontzilting installatie te maximaliseren. Het doel van de optimalisatie is het probleem op te lossen via de uitbreiding van de productiecapaciteit.

Acknowledgments

This study has been stimulated by inspiration of various people who have constantly encouraged me throughout my research. My gratitude goes to my family: my mother and father who were always an inspiration to me in my life hoping to achieve a fraction of what they have done with their hard work and dedication. I can not thank you enough for all what you have done and still doing for me. To my sisters Marwa and Youssra, without their help, I would not have been able to make it; many are the times when they put my interest before their own just to help me keep going. My thanks goes to my father in law, Dr. Ibrahim Badawi who introduced me to UNESCO-IHE and was the reason I started this journey in the first place; to my mother in law, who encouraged me at times when I thought of dropping everything and letting go; to my brother in law, whose passion in pursuing academic excellence was always a motivation.

My deepest thanks to my beloved husband, Karim, who was always there for me giving me love and support and making me believe in myself and in my ability to conclude this demanding job. You are one of a kind. My thanks to my precious little ones Youssef and Yasmine whose simple kind words of love was all that I needed at times when things got tough.

My sincere thanks to my promoter Professor Pieter van der Zaag for his guidance and assistance at times when I knew he was extremely busy. I wish to thank my supervisor Dr. Elisabeth von Münch who during the period she stayed at UNESCO-IHE was a continuous help and support. My sincere thanks also goes to Professor Emad Imam at The American University in Cairo for guiding me throughout the thesis and helping to set a workflow which is interesting and beneficial to the case at hand.

I would like to thank Dr. Amaury Tilmant for taking the time and effort to review some of my papers and giving me essential feedback. My sincere thanks to Dr. Maria Kennedy, Mrs. Susan Graas, Mr. Daniel Schotanus, Mr. Wim Spaans for their feedback on bits and pieces of this research. I would like to thank Jolanda Boots who was always present to answer our countless questions making sure everything is progressing smoothly during the period of our research.

A main part of my thesis depended on data collection. Since published work is very limited on the case study region, I had my doubts at the beginning that I could find the resources and key people who would help and be willing to spend time and effort and share information. I was proven wrong through the pool of generous people who graciously helped me amidst their busy and overcharged schedule.

I would like to greatly acknowledge the help and support I got from Professor dr. Magdy Abou Rayan, Mansoura University, and Dr. Ibrahim Khaled, Sinai Development Authority, who did not spare an effort in introducing me to relevant people and giving out any information I needed for the research. My sincere thanks to Mr. Mohamed Abd Al Latif for his patience in answering all my questions and welcoming me each time I went to Sharm.

This thesis would not have been possible without the great help and assistance of Mr. Zaki Girges, CEO of Ridgewood Egypt. He entrusted me with confidential business information which has enormously contributed to the value of this research. Amidst his busy schedule, he would always welcome me in his office sharing his working experience in the case study area and continuously encouraging me all along the way. I would also like to extend my special thanks to the rest of Ridgewood team: Mr. Mostafa Daoud, Mr. Haytham Rafee, Mr. Adel El Shenhab and Mrs. Lobna Refaat. I was considered as part of the team bothering them for hours for long weeks gathering all the information I needed and benefiting from their vast working experience.

My thanks to Dr. Azza Hafez, Research professor at the National Research Centre, Egypt and Dr. Samir El Manharaway for their generous help and support especially during the early difficult stages of the research. I would like to thank Dr. Mushtaque Ahmed, Associate Professor at Sultan Qaboos University, Oman; and Dr. Hani El Nokrashy, Nokrashy Engineering in Germany, for sharing his valuable thoughts on parts of the research.

I would like to extend my special thanks to Dr. Magdy Saleh, Managing Director of Global Energy; Mr. Mostafa El Moandy, Financial manager of Fantazia hotel; Mr. Mohamed Moharam, General manager of South Sinai water company; Mr. Khaled Al Abbassy, Director of Engineering at Conrad hotel; Dr. Ahmed Said; Mr. Arthur Gomes, Director of Engineering at Jolie Ville Resort and Casino; Mr. Mohamed Abd Al kader, Al Montaza water company; Mr. Yasser Ibrahim, Mr. Seif El Messih, Orascom, Mr. Tarek Waly and Mrs. Amel Azab, UNESCO-IHE, all for sharing information and enriching this research.

Curriculum Vitae

Aya Lamei was born on the 15th of October 1975 in Cairo, Egypt. She joined the American University in Cairo in 1991 where she enrolled in a Bachelor of Science degree majoring in Construction Engineering and graduated in 1996. In 1996, she was employed at the Egyptian Construction and Trading Company as a project manager. In the same year, she started an MSc program at the American university in Cairo, majoring in Environmental Engineering. She graduated in 2001.

She worked at the American University in Cairo both as a research and teaching assistant in 2004 and 2007 respectively.

She enrolled in a PhD program at UNESCO-IHE Delft in 2005. Her study was funded by the Netherlands Fellowship Programme (Nuffic). Her PhD was a sandwich construction programme whereby she conducted field work in Sharm El Sheikh, Egypt under the supervision of Professor, dr. Emad Imam, at the American University in Cairo. The consultation, course work and part of the thesis writing took place at UNESCO-IHE.

She is currently employed in Schlumberger Water and Carbon Services, Texas, United States of America as a business development manager.